The Cretaceous/Tertiary boundary interval, Raton Basin, Colorado and New Mexico, and its content of shock-metamorphosed minerals; Evidence relevant to the K/T boundary impact-extinction theory

Glen A. Izett
U.S. Geological Survey
M.S. 913, Box 25046
Denver Federal Center
Denver, Colorado 80225

THE GEOLOGICAL SOCIETY OF AMERICA · 1888 ·

SPECIAL PAPER

249

Published by The Geological Society of America, Inc.
3300 Penrose Place, P.O. Box 9140, Boulder, Colorado 80301

GSA Books Science Editor Richard A. Hoppin

Printed in U.S.A.

Library of Congress Cataloging-in-Publication Data
Izett, Glen Arthur, 1928– .
The Cretaceous/Tertiary boundary interval, Raton Basin, Colorado and New Mexico, and its content of shock-metamorphosed minerals : evidence relevant to the K/T boundary impact-extinction theory / Glen A. Izett.
p. cm.—(Special paper / Geological Society of America ; 249)
Includes bibliographical references.
ISBN 0-8137-2249-7
1. Cretaceous/Tertiary boundary—Raton Basin (Colo. and N.M.) 2. Geology—Raton Basin (Colo. and N.M.) 3. Mineralogy—Raton Basin (Colo. and N.M.) I. Title. II. Series: Special papers (Geological Society of America) ; 249.
QE689.I94 1990
551.7'7'0978896—dc20 90-3737
CIP

Cover photos: Front, K/T interval from the Clear Creek North site, Colorado (Continental rocks in western United States). Back, K/T interval from the Stevns Klint, Denmark, site (Marine rocks in Europe).

Contents

Acknowledgments

The writer is indebted to many U.S. Geological Survey colleagues for sharing information they have gathered concerning K/T boundary rocks, including R. F. Fleming, E. E. Foord, P. J. Modreski, D. J. Nichols, and J. A. Wolfe. Special thanks are given to C. L. Pillmore who convinced me in late 1984 to investigate K/T boundary rocks in the Raton basin. He kindly guided me to K/T boundary sections that he had discovered and studied. He patiently engaged in seemingly endless discussions concerning the origin of K/T boundary interval rocks.

E.C.T. Chao was a constant source of pertinent information and advice relative to shock-metamorphic effects in silicate minerals. Z. E. Peterman, M. J. Kunk, and J. F. Sutter made available unpublished isotopic data they acquired. B. F. Bohor provided information and samples of K/T boundary interval rocks in his collection. E. M. Shoemaker kindly sent me an oriented sample of rock that spans the K/T boundary at Stevns Klint, Denmark. H. T. Millard, Jr., skillfully made radiochemical INAA analyses of samples of K/T boundary rocks that I had collected. I also thank C. G. Whitney and R. A. Sheppard who made x-ray diffraction analyses of samples of claystone in the K/T boundary interval. J. Nishi graciously advised and aided me during SEM study of K/T boundary rocks in his laboratory. N. H. Bostick provided useful information about the nature of the coaly material in the boundary claystone and impact layer. I am indebted to F. E. Lichte for making chemical analyses of K/T boundary rocks and minerals as part of his development of a new analytical chemical system (laser-ablation inductively coupled mass spectrometer).

R. R. Anderson, of the Iowa Department of Natural Resources, sent me core from the Manson 2A core hole and shared unpublished information pertaining to the Manson, Iowa, impact structure. He also reviewed an earlier version of this manuscript and made many beneficial comments. I thank B. Schmitz, of the University of Göteborg and H. Michel and F. Asaro of the Lawrence Berkeley Laboratory of the University of California; they determined the amount of iridium in samples that Schmitz and I collected at the Carmel K/T boundary site near Madrid, Colorado. W. C. Elliott of Michigan State University generously provided me with a sample of the K/T boundary interval he collected at Stevns Klint, Denmark.

An earlier version of this paper constituted my Ph.D. dissertation at the University of Alaska, and as such, it was critically read by my committee (D. L. Turner, chairman, J. E. Beget, J. Kienle, C. W. Naeser, S. E. Swanson, and D. M. Triplehorn). In addition, it was also reviewed by H. T. Millard, Jr., N. D. Naeser, and C. L. Pillmore. I appreciated all their thoughtful comments, most of which improved the paper.

Geological Society of America
Special Paper 249
1990

ABSTRACT

At about 20 sites in the Raton basin of Colorado and New Mexico and at least 10 sites in Wyoming, Montana, and western Canada, a pair of claystone units, an iridium (Ir) abundance anomaly, and a concentration of shock-metamorphosed minerals mark the palynological Cretaceous/Tertiary (K/T) boundary. The lower unit, the K/T boundary claystone bed, is generally 1 to 2 cm thick; the upper unit, the K/T boundary impact layer is, on average, 5 mm thick. This couplet is overlain by a coal bed 1 to 16 cm thick.

The boundary claystone in the Raton basin consists mainly of kaolinite and small amounts of illite/smectite (I/S) mixed-layer clay; but to the north at some sites in Wyoming, Montana, and Canada, it is more smectitic. Where the boundary claystone is kaolinite rich, it is similar to tonstein layers in coal. Typically, the claystone contains deformed vitrinite laminae, root-like structures, and plant impressions. The microscopic texture of the claystone is a polygonal framework of kaolinite filled with micrometer-size kaolinite spherules. The claystone is fragmental; at some sites, particularly in Wyoming, it contains centimeter-size kaolinite pellets, some of which contain carbonaceous plant material and millimeter-size goyazite spherules. The suite of trace elements in the claystone is, in general, similar to the suite of trace elements in tonsteins and to that in average North American shale, except that the boundary claystone contains anomalous amounts of platinum-group elements, including Ir (0.07 to 0.32 ppb).

The K/T boundary impact layer also consists of kaolinite and considerably more (I/S) mixed-layer clay than the boundary claystone. Commonly it is 3 to 8 mm thick, microlaminated, and contains planar laminae of vitrinite and ubiquitous kaolinite barley-shaped pellets similar to the "graupen" of tonsteins. The microscopic texture of the claystone as seen with a scanning electron microscope (SEM) generally consists of an open wavy framework similar to that of smectite and (I/S) mixed-layer clay. This texture is significantly different than the microspherulitic texture of the underlying K/T boundary claystone. The contact between the impact layer and boundary claystone is generally sharp, and locally it may be a diastem.

The impact layer and boundary claystone are similar chemically, but the former has slightly more Fe, K, Ba, Cr, Co, Li, V, and Zn than the latter. Iridium is most abundant in the impact layer (1.2 to 14.6 ppb); however, anomalous amounts of this element are found in carbonaceous shale and coal below or above the impact layer (0.1 to 2.0 ppb).

The fact that the boundary claystone and impact layer make up a regionally extensive unit suggests they are composed of primary air-fall impact or volcanic material. However, mineralogic and chemical evidence indicates that the K/T boundary claystone is not composed of altered impact ejecta because it essentially lacks shock-metamorphosed minerals and contains only a minor Ir anomaly. Moreover, it contains only a few parts per million of Ni; this is inconsistent with the idea that it is composed of altered impact material. Observational and chemical evidence disclosed that the shocked minerals and Pt-group elements, including Ir, are the only impact-related components in the impact layer. Seemingly, the boundary claystone and impact layers are not of volcanic origin because they essentially lack a *coherent assemblage* of volcanic crystals.

In the Raton basin, the boundary claystone, but not the impact layer, contains two types of spherules: kaolinite and goyazite (hydrous aluminum phosphate). Spherules composed of two Fe-rich minerals—jarosite or goethite—occur in the boundary claystone and impact layer. Kaolinite spherules are always solid and consist of a microspherulitic core encased in a thin shell of columnar kaolinite, whereas goyazite spherules are generally hollow and consist of a microlaminated colloform shell. Kaolinite spherules are large (0.1 to 0.5 mm) and relatively common (0.1 percent); in contrast,

goyazite spherules are small (<0.1 mm) and rare (<0.01 percent). Kaolinite and goyazite spherules are dispersed widely in the claystone, but inexplicably they also occur in clusters. At two Wyoming sites, the boundary claystone contains abundant (as much as 30 percent) large (0.9 mm) goyazite spherules. Evidence indicates that the spherules formed by authigenesis.

Although the impact layer is mainly composed of clay, it contains a small amount, as much as 2 percent, of clastic mineral grains. About 50 percent of the grains are quartz, and about 50 percent of these contain multiple sets of shock lamellae. Quartzite and metaquartzite constitute 26 percent of the assemblage of clastic grains in the impact layer, and 30 percent of these contain multiple sets of shock lamellae. Clearly the source of the quartzite and metaquartzite grains was chiefly sedimentary, metasedimentary, and metamorphic rocks and not volcanic rocks as some have suggested. Unshocked grains of chert and chalcedony compose from 8 to 46 percent of the assemblage. Grains of shocked feldspar (oligoclase and microcline) and granite-like mixtures of quartz and feldspar are rare. The abundance of unshocked quartzite, metaquartzite, and chert in the impact layer as compared to their paucity in underlying Cretaceous rocks suggests that most of these grains were derived from continental supracrustal impact target rocks and are not locally derived material.

Shock-metamorphosed mineral grains are larger (mean = 0.15 to 0.20 mm) at North American K/T boundary sites relative to boundary sites elsewhere in the world (mean = 0.09 mm). Large (0.25 to 0.64 mm) grains are found at all Western Interior North American K/T boundary sites. Moreover, shocked minerals are several orders of magnitude more abundant at these sites as compared to elsewhere in the world. This information indicates unambiguously that the K/T impact occurred in North America.

The Manson, Iowa, impact structure may be the K/T impact site because of the mineralogic similarity of Manson area subsurface rocks and shocked K/T boundary minerals, the large size (36 km) of the structure, the compatible isotopic age (~66 Ma) of shocked granitic rock from the Manson structure and the K/T boundary, and the proximity of the Manson structure to North American K/T boundary sites that contain relatively abundant and large shock-metamorphosed minerals.

INTRODUCTION

In 1979 and 1980, a team of scientists headed by L. W. Alvarez of the University of California reported finding trace amounts of the platinum-group metal iridium in a 1-cm-thick marine claystone bed at the paleontologic Cretaceous/Tertiary (K/T) boundary near Gubbio, Italy (Alvarez and others, 1979a, 1979b, 1980). The peak amount of iridium they measured, 9.1 parts per billion (ppb), in the claystone bed was about 30 times greater than that in subjacent Cretaceous and superjacent Paleocene limestone beds. Their discovery was significant because iridium generally occurs in minute amounts, less than 0.1 ppb, in most rocks that compose the Earth's crust (Palme, 1982, Fig. 1). It is present in much greater concentration (500 to 1,000 ppb) in primitive cosmic materials such as chondritic meteorites and presumably in the Earth's core. They also reported finding anomalously large amounts of iridium in marine sedimentary rocks at the K/T boundary at such widely scattered sites as Denmark and New Zealand. Their purpose in analyzing the claystone for the siderophile element iridium was to determine, if possible, the amount of time that elapsed during the deposition of the 1-cm-thick claystone (Alvarez and others, 1980, p. 1096). They apparently were aware of the suggestion of Pettersson and Rotschi (1952, p. 90) that anomalously large amounts of platinum-group metals, including iridium and osmium, might be found in deep ocean-floor sediments owing to the influx of meteoritic dust into the oceans. Subsequently, Barker and Anders (1968, p. 627) measured trace amounts of osmium and iridium in sea-floor sediments by the instrumental neutron activation method and set an upper limit for the accumulation rate for these elements. If the influx rate of meteoritic dust into the oceans were constant, it would provide a method for calculating the depositional rate of the sediments.

Because the iridium anomaly coincides exactly with the classic mass extinction of animals that occurred at the close of the Cretaceous (Loeblich and Tappan, 1964; Raup and Sepkoski, 1982), the Alvarez team proposed a theory, here called the "impact-extinction theory," linking these two observations. They speculated that a 10 ± 4-km-size Apollo asteroid struck Earth and formed a crater 200 km in diameter. The Alvarez team suggested that the impacting body was an asteroid rather than a comet. I also refer to the extraterrestrial object as an asteroid, but for discussion purposes only, because the exact nature of the impacting body is unknown. They proposed that an enormous amount of fine-grained target material (60 times the asteroid mass) was injected into the stratosphere, and some of the dust-size impact ejecta stayed aloft and blocked sunlight for several years. As a result, global darkness followed, the level of photosynthesis markedly decreased, and the food chain collapsed, resulting in the extinction of about half of Cretaceous life forms, including the dinosaurs. This extinction event has been used to define the end of Cretaceous time. I prefer, for reasons given in one of the last sections of this paper, to use 66 Ma for the isotopic age of the K/T boundary, rather than the commonly cited 65 Ma.

The discovery of the iridium anomaly in K/T boundary rocks by the Alvarez team prompted a search for evidence that would test their impact-extinction theory. Considerable debate has arisen, especially in regard to the nature of the killing mechanism, the fabric and timing of the mass-extinction event, the composition and size of the extraterrestrial body, and the location and mechanics of formation of the impact crater. There has been an enormous growth in the number of papers reporting the results of research since 1980. Reports have been published by astronomers, biologists, biostratigraphers, geochemists, geologists, geophysicists, oceanographers, mineralogists, paleontologists, physicists, planetary geologists, and experts in cratering mechanics. So many papers have been written that no attempt is made here to list all of the numerous reports generated as a result of the Alvarez theory. Many references can be found in papers published as a result of a conference titled "Large Body Impacts and Terrestrial Evolution—Geological, Climatological, and Biological Implications" held at Snowbird, Utah, in 1981 (Silver and Schultz, 1982). Paleontological information can be found in the proceedings of (1) a symposium—"Cretaceous/Tertiary Boundary Events" (Birkelund and Bromley, 1979); (2) a workshop—"Cretaceous-Tertiary Extinctions and Possible Terrestrial and Extraterrestrial Causes" (Russell and Rice, 1982); (3) a book—*Catastrophes and Earth History* (Berggren and Van Couvering, 1984); and (4) the abstracts and papers of a conference on mass-extinction events held in Bilbao, Spain (Lamolda and Cearreta, 1987). Recent developments can be found in the abstracts of papers delivered at a K/T boundary conference held at Snowbird, Utah, in 1988.

An important prediction of the impact-extinction theory is that an iridium anomaly should be found globally at complete K/T boundary stratigraphic sections. This prediction has been elegantly confirmed because more than 90 boundary sections containing an abundance anomaly of iridium (and other noble metals) have been reported (Alvarez and others, 1980; Ganapathy, 1980; Smit and Hertogen, 1980; Kyte and others, 1980; Alvarez and others, 1982; Smit and Kate, 1982; Luck and Turekian, 1983; Brooks and others, 1984; Gilmore and others, 1984; Kastner and others, 1984; Orth and others, 1984; Alvarez and Montanari, 1985; Kyte and others, 1985; Lerbekmo and St. Louis, 1985; Smit and Romein, 1985; Alekseev and others, 1986; Lichte and others, 1986; Bohor and others, 1987; Izett, 1987c; Pillmore and Flores, 1987; Strong and others, 1987; Johnson and others, 1989; Kaenel and others, 1989; Rosenfeld and others, 1989). Be this as it may, the origin of a geochemical anomaly, such as the K/T boundary iridium spike, can be open to multiple interpretations, and it is not surprising that a few earth scientists have offered alternative ideas to explain its origin. Officer and Drake (1985), McLean (1985), and Officer and others (1987) proposed that the iridium is of terrestrial volcanic rather than of extraterrestrial impact origin, but Alvarez (1986) argued persuasively in defense of the impact origin of the iridium.

Although the origin of the K/T boundary iridium anomaly is controversial, there is directly observable *physical evidence* indicating that a large extraterrestrial object struck Earth at the close of the Cretaceous, ~66 Ma. In K/T boundary rocks in the Hell Creek area of eastern Montana, Bohor and others (1984) discovered quartz grains that contain an unusual type of microstructure consisting of multiple intersecting sets of closely spaced planar lamellae. They concluded that the lamellae are identical to those formed in quartz that has undergone a high strain rate during hypervelocity shock in laboratory experiments and in quartz and other minerals at confirmed meteor impact sites. These multiple sets of planar microstructures also are found in silicate minerals of rocks at nuclear and high-yield chemical explosion sites. For brevity, I use the term "shocked" rather than "shock metamorphosed," and "shock lamellae" rather than "multiple intersecting sets of planar lamellae" at some places in this paper.

Following the discovery by Bohor and others (1984) of shock-metamorphosed quartz grains in K/T boundary rocks in Montana, Izett and Pillmore (1985a, 1985b) found not only grains of shock-metamorphosed quartz, but also rare grains of shock-metamorphosed oligoclase and a potassium-rich feldspar (microcline) in K/T boundary rocks in the Raton basin of Colorado and New Mexico. In addition, shock-metamorphosed mineral grains have been found in K/T boundary rocks at other western North America sites, including those in Wyoming, North Dakota, Alberta, and Saskatchewan (Alexopoulos and others, 1987; Bohor and others, 1987; Wolfe and Izett, 1987; Grieve and Alexopoulos, 1988; Johnson and others, 1989; Nichols and others, 1990). On a global scale, shock-metamorphosed mineral grains have been found in K/T boundary rocks in Italy, Denmark, Spain, New Zealand, and the north Pacific Ocean by Bohor and Izett (1986) and Izett (1987a, 1987c, 1988); in Austria by Preisinger and others (1986); and in Asia by Badjukov and others (1986). These discoveries provide the principal *physical evidence* supporting the impact component of the impact-extinction theory of the Alvarez team.

Concerning the "extinction" component of the impact-extinction theory, many ideas have been advanced to explain mass-extinction events such as the classic event that defines the end of the Cretaceous. For example, Laubenfels (1956) suggested that the impact of a large meteorite caused the extinction of the dinosaurs. McLaren (1970) and Urey (1973) proposed an extraterrestrial cause for mass-extinction events in the Phanerozoic, but unlike the Alvarez team, they did not present any physical or chemical evidence in support of their theory.

The proposal by the Alvarez team that the impact had a profound effect on Cretaceous life forms has not been accepted by many scientists, especially some paleontologists who have specialized in the study of marine fossils (see review by Hallam, 1987). Many paleontologists believe that the K/T extinction event was only a part of a broad pattern of extinctions that began in the Late Cretaceous in response to lowering of sea level. Surely, a complete understanding of the effects of the impact on marine Cretaceous life forms will take more biostratigraphic data than are now available. More detailed biostratigraphic studies are needed, such as those of Strong and others (1987), Johansen (1987), Keller (1988, 1989), Smit and others (1988), and Ward and MacLeod (1988).

The influence of the K/T asteroid impact on land plants was minimal according to Hickey (1984, p. 303). Tschudy and others (1984) claimed that a massive destruction of vegetation and the extinction of a few plant species took place as a result of some catastrophic event at the close of the Cretaceous Period. Tschudy and Tschudy (1986, p. 667) stated that "No massive total extinction of plant genera at the end of the Cretaceous can be seen from the palynological record." Nevertheless, they maintained that the flora of the Western Interior of North America underwent a profound shock at the exact K/T boundary. Apparently this floral disruption was limited to the Western Interior of North America. Johnson and Hickey (1988, p. 87) concluded that the patterns of floral change across the K/T boundary were associated mainly with long-term environmental processes and only partly with asteroid impact processes. In western North Dakota, Johnson and others (1989) recently determined that a major palynofloral and megafloral extinction event occurs at the lithologically defined K/T boundary. The floral extinction event is at the same stratigraphic level as an iridium anomaly and a concentration of shock-metamorphosed quartz grains. They speculated that the extinction event was caused by the impact of an extraterrestrial body.

The cause for the extinction of the dinosaurs has been of renewed interest in the last decade because Alvarez and others (1980) suggested that the impact of an asteroid caused their demise. Relative to the extinction of dinosaurs, Schopf (1982, p. 416) stated that "The major point of this and the previous paragraphs is that dinosaurs appear to be last known from beds older than latest Maastrichtian in all places except in northern Europe and in the Western Interior of North America."

Although Alvarez and others (1980) proposed that the K/T impact caused the extinction of the dinosaurs, their fossil remains occur no higher than 1 m below the level of the K/T boundary iridium anomaly and shocked minerals. The reason for this 1-m stratigraphic gap is unknown. Archibald and Clemens (1982, p. 380) observed that dinosaur remains disappear from the stratigraphic section in the Hell Creek area of Montana about 3 m below a coal bed that marks the K/T boundary. In Alberta, Canada, Lerbekmo and others (1979) and Lerbekmo and St. Louis (1985) reported that the stratigraphically highest dinosaur remains are 4.5 m below the paleontologic K/T boundary. Bohor and others (1987) found a large fragmented dinosaur bone in a sandstone 1 m below the K/T boundary claystone at the Dogie Creek, Wyoming, site. The stratigraphically highest record of dinosaur fossil remains in the Hell Creek Formation of western North Dakota is 1 m below the K/T boundary (Johnson and others, 1989). I recently collected dinosaur bone fragments (float) from a stratigraphic level about 6 m below the K/T boundary claystone at the Sussex, Wyoming, K/T boundary site of Nichols and others (1990).

Sloan and others (1986) argued that the extinction of the dinosaurs was a gradual process that began 7 m.y. before the end of the Cretaceous. Recently, Rigby and others (1987) found dinosaur teeth within alleged Paleocene channel-fill deposits in the Hell Creek area of Montana. The age assignment of the rocks containing the dinosaur teeth was based on palynomorphs said to be of Paleocene age. Apparently, neither the K/T boundary claystone, shock-metamorphosed minerals, nor an iridium abundance anomaly have been found at their specific fossil sites.

PURPOSE OF THE STUDY

Descriptions of some aspects of the stratigraphic location and mineralogic features of shock-metamorphosed minerals in K/T boundary rocks in western North America have been reported (Izett, 1987c, 1988; Bohor and others, 1987). However, a complete description of important stratigraphic and mineralogic information has not been reported fully, and this information can be used to constrain models for the K/T boundary impact event. Many papers have been written on K/T boundary subjects; some were written unconstrained by stratigraphic and mineralogic data. It was, therefore, the purpose of this study to (1) describe the sedimentary rocks that constitute the K/T boundary interval, chiefly in the Raton basin of Colorado and New Mexico, but also at other complete K/T boudnary sections in western North America; (2) provide petrographic information pertinent to the shock-metamorphosed minerals in the K/T boundary impact layer; (3) present mineralogic evidence that indicates the minerals containing shock lamellae are of impact and not volcanic origin; and (4) suggest that the K/T boundary asteroid struck Earth in North America in an area underlain by continental crust composed of sedimentary, metasedimentary, metamorphic, and granitic rock, such as the area of the Manson, Iowa, impact structure.

It is important to keep in mind that the size, composition, density, velocity, and strike angle of the K/T asteroid will never be known with assurance. Thus, the intensity and magnitude of the K/T boundary impact and impact processes, including the transportation and deposition of the impact ejecta, are largely speculative. Moreover, until recently, distal air-fall (or reworked air-fall) impact-ejecta deposits, other than those described in this study, have rarely been reported in the geological literature. A few recent papers have described such distal impact deposits. Gostin and others (1986, 1989) found, studied, and described a deposit of air-fall impact material in Proterozoic sedimentary rocks of South Australia, and Glass (1987) and Thein (1987) described a layer of upper Eocene coarse-grained glassy impact ejecta at the base of an 8- to 12-cm-thick bed in the core of DSDP site 612 that lies on the continental slope off the New Jersey coast. Because distal impact deposits of this type, including the K/T boundary deposits, are not fully understood, some of the interpretations and conclusions presented in this paper are speculative, controversial, and provocative.

K/T BOUNDARY INTERVAL ROCKS IN THE RATON BASIN

In the Raton basin, the K/T boundary occurs in a sequence of fine-grained, continental sedimentary rocks deposited in fluvial, pond, and swamp environments following the retreat of the Western Interior Cretaceous seaway. The uppermost Cretaceous and lowermost Tertiary rocks include, from oldest to youngest, the marine Pierre Shale and Trinidad Sandstone and the continental Vermejo and Raton Formations. The Pierre is about 600 m thick, and the Trinidad about 30 m thick. The Vermejo Formation, which varies from 0 to 115 m in thickness, overlies the Trinidad and consists of alternating beds of sandstone, siltstone, carbonaceous shale, and mudstone. Coal beds as much as 4 m thick occur mainly in the lower and upper parts of the formation. The Vermejo is overlain unconformably by the Raton Formation of Cretaceous and Paleocene age, and the contact is marked at some places by a conglomeratic sandstone at the base of the Raton Formation. The paleontologic K/T boundary occurs in the lower part of the Raton Formation near the top of the lower coal zone in a sequence of rocks similar to those of the underlying Vermejo Formation. The very generalized stratigraphic framework outlined above is a much-condensed version of descriptions given by Pillmore and Flores (1987).

The boundary interval is composed of a pair of thin claystone units that were informally called the " K/T boundary claystone bed" and " K/T boundary impact layer" by Izett (1987c, 1988). This couplet of claystone layers is directly overlain at most sites by a thin, persistent coal bed, informally called the " K/T boundary coal bed" (Fig. 1). Because the impact layer always contains a concentration of shock-metamorphosed mineral grains and generally registers the peak iridium abundance anomaly in the K/T boundary interval, I called it colloquially the "magic layer." I did this to emphasizae to my coworkers at the U.S. Geological Survey that shock-metamorphosed minerals and the peak iridium abundance anomaly abruptly appear in this unit and not in the underlying boundary claystone. I did not intend for this term to be used as a stratigraphic name in a formal publication, but unfortunately Bohor and others (1987) used the term in a stratigraphic sense. In this paper, I have again termed it, for discussion purposes only, the K/T boundary impact layer. The use of the term "impact layer," is somewhat misleading and suggests that it is composed entirely of impact-derived material. I intend to show in subsequent sections of this paper that quite the opposite is true; the unit contains only a small amount of impact-generated material.

At some sites within the Raton basin such as Sugarite, Madrid East, and Carmel (see Fig. 2 for locations), the boundary claystone and impact layer were deposited in a coal swamp because they are interbedded with coal or impure coal. But at other sites (Raton, Clear Creek North, and Berwind), these units were deposited just prior to the inception of a coal-forming event because they occur above noncarbonaceous mudstone or carbonaceous shale that formed in fluvial and pond environments and below the boundary coal bed that formed in a coal swamp. At K/T boundary sections in Wyoming and Montana and farther north in Alberta and Saskatchewan, Canada, the boundary claystone bed and impact layer also were deposited just prior to the onset of a coal-forming event. At the Pyramid Butte, North Dakota, site of Johnson and others (1989), an iridium anomaly and shock-metamorphosed quartz grains occur in a bioturbated interval that formed after a coal-forming event. This interval contains thin, diffuse lenses of boundary clay lithology, but the impact layer was not identified.

Because the boundary claystone bed, impact layer, and superjacent units consist of claystone, mudstone, carbonaceous shale, and coal deposited in a low-energy depositional regime, they are especially well suited for study of fine-scale sedimentary structures that record in detail K/T boundary depositional events. Moreover, the lithologic succession that contains the boundary is similar at each site in the Raton basin; this is surprising considering that the rocks were deposited in continental sedimentary environments commonly noted for their lithologic heterogeneity. More surprising, the same sequence of K/T boundary units that occurs in the Raton basin was identified by me in 1986 and 1987 in samples from the Lance Creek, Wyoming, area (Bohor and others, 1987), in the western Powder River basin, Wyoming (Wolfe and Izett, 1987; Nichols and other, 1990), and at Brownie Butte in the Hell Creek area of Montana (Bohor and others, 1984). From outcrops farther north, I identified these same units in samples from the Morgan Creek and Frenchman River sites in Saskatchewan (Nichols and others, 1986; Lerbekmo and others, 1987) and from the Red Deer Valley site in Alberta (Lerbekmo and St. Louis, 1985).

At all of these Western Interior North American sites, there is no record in the fine-grained sedimentary rocks that directly overlie the K/T boundary of any pronounced disruption of the sedimentological regime. More specifically, there are no extensive conglomeratic channel sandstone beds or coarse-grained sediments directly above the K/T boundary impact bed. Surely such deposits would have formed soon after the K/T impact if a major deforestation event had occurred as a result of speculative global wildfires or acid-rain fallout (Wolbach and others, 1985; Wolbach and others, 1988; Prinn and Fegley, 1987). This speculative deforestation event would have been most pronounced if the K/T asteroid struck North America as maintained herein. The lack of physical evidence in the form of a profound stratigraphic disruption directly above the K/T impact layer implies that there were no catastrophic environmental events associated with the K/T impact in the Western Interior of North America.

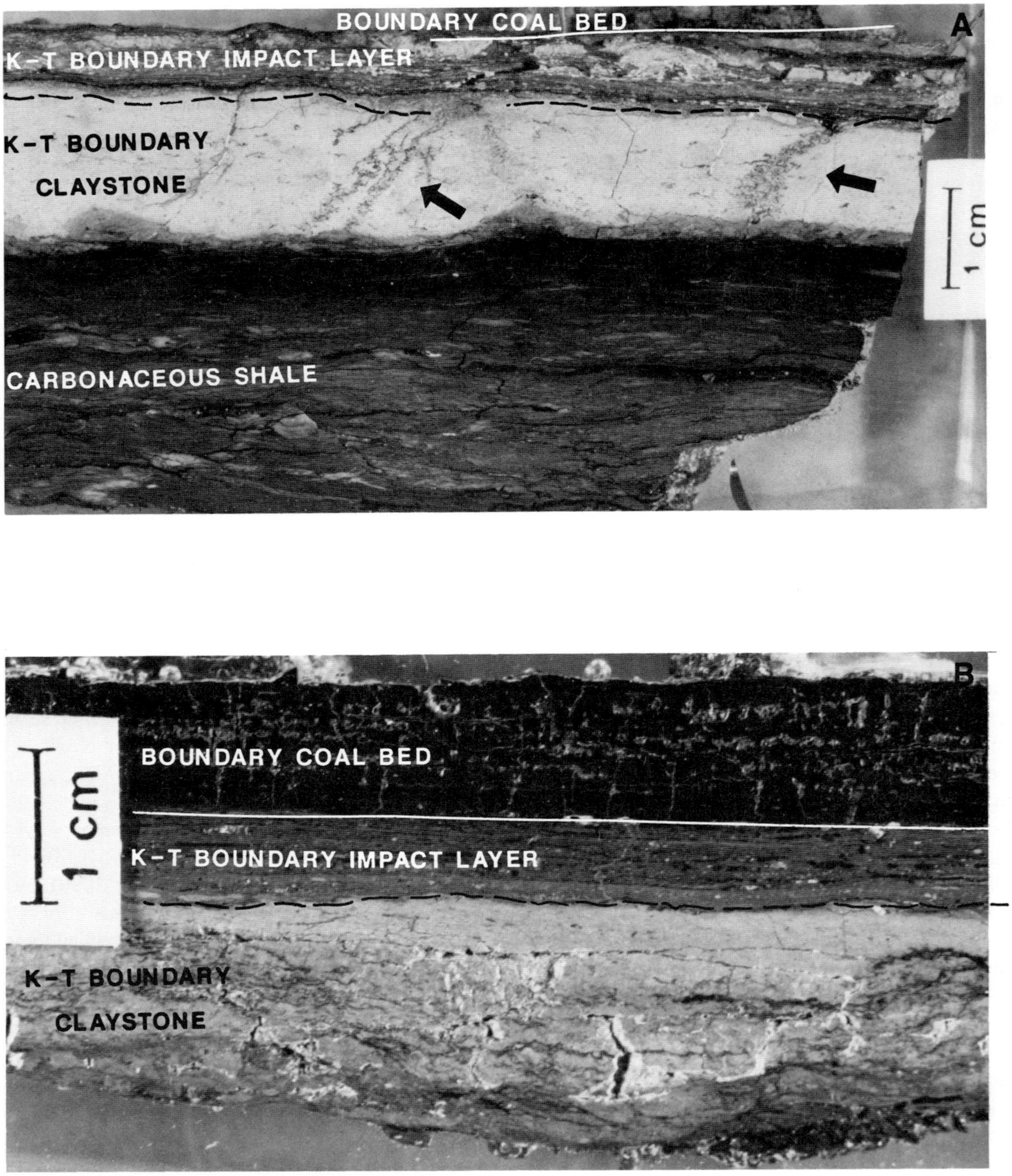

Figure 1. Polished slabs of the K/T boundary interval from the Raton basin, Colorado. The K/T boundary interval consists of carbonaceous shale, the boundary claystone, the impact layer, and the boundary coal bed. A, specimen from the Clear Creek North site, 10 km south of Trinidad, Colorado. Two prominent carbonaceous subvertical rootlike structures (shown at arrows) cut the boundary claystone but not the impact layer. Clastic quartz and other mineral grains are found along such structures but not in surrounding claystone. B, specimen from the Berwind Canyon K/T boundary site, about 20 km northwest of Trinidad, Colorado. Carbonaceous shale below boundary claystone lost during preparation.

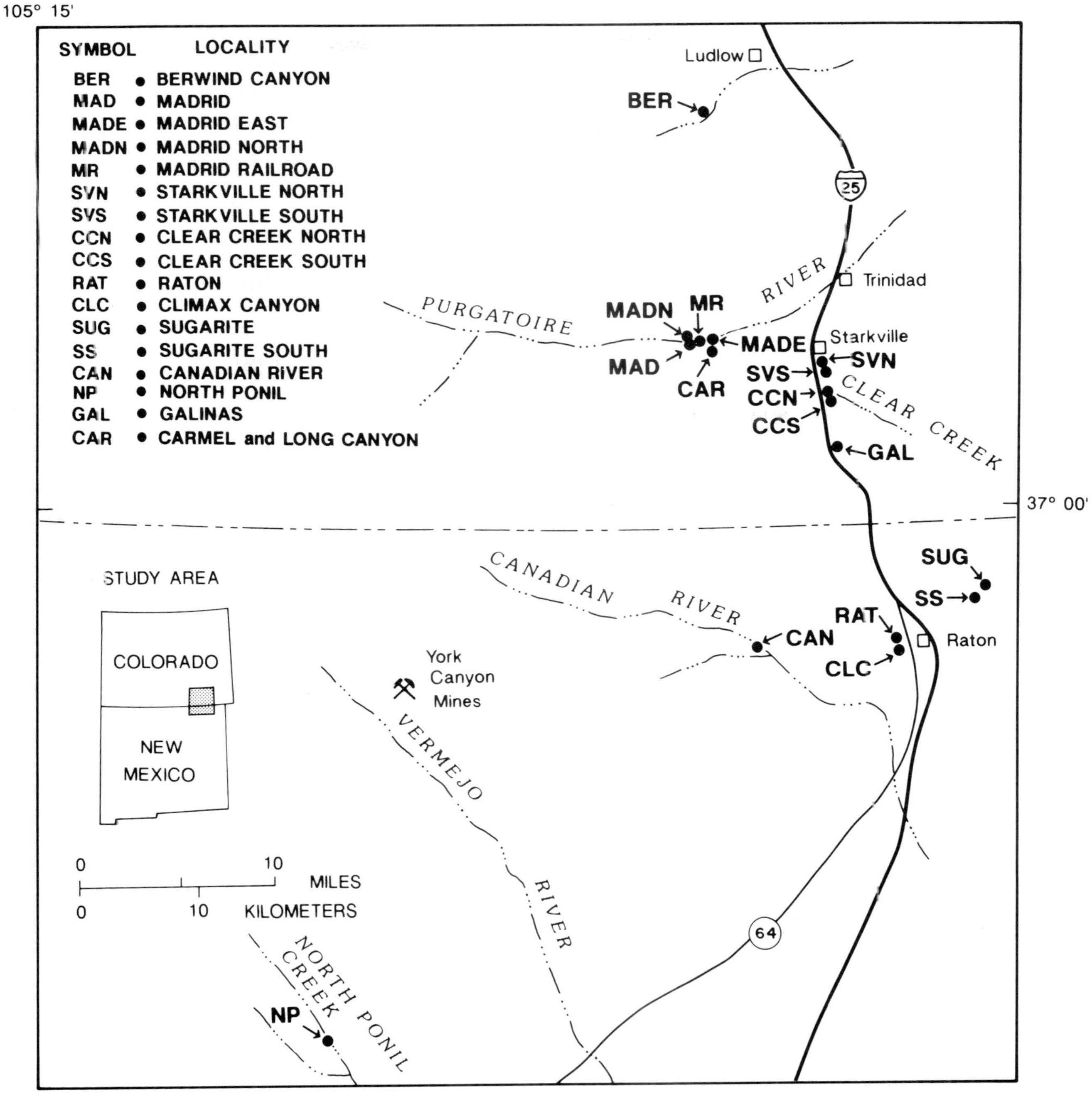

Figure 2. Index map for the Raton basin K/T boundary sections mentioned in the text. A list of abbreviations used in Tables 1 through 9 for these sections is given. Modified from Pillmore and Flores (1987).

ROCKS DIRECTLY UNDERLYING THE K/T BOUNDARY CLAYSTONE

The rocks of the Raton Formation that directly underlie the K/T boundary claystone in the Raton basin are composed mainly of several tens of centimeters of mudstone, carbonaceous shale, coal, impure coal, or sandstone. Of these, carbonaceous shale is the most common. The clay minerals that form the mudstone and shale are composed of ordered and randomly ordered illite/smectite (I/S) mixed-layer clay, illite, and kaolinite (Pollastro and Pillmore, 1987; C. G. Whitney, U.S. Geological Survey, written communication, 1986).

At many places, the mudstone and carbonaceous shale are dark brownish gray and carbonaceous; at other places, they are light brownish gray and less carbonaceous. For example, at the Raton, New Mexico, site, light gray mudstone several tens of centimeters thick lies below the K/T boundary claystone bed. Fragments of macerated plant material and leaf impressions occur on many bedding surfaces of the carbonaceous shale. Locally, the shale and mudstone contain rounded blebs of cellular fusinite as much as 1.0 mm in diameter.

On a microscopic scale, the bedding of the carbonaceous shale consists of planar laminations whose presence is emphasized by discontinuous laminae of coaly carbonaceous material that is probably vitrinite (N. H. Bostick, U.S. Geological Survey, written communication, 1986). Discontinuous 3.0-mm-thick laminae and lenses of a jarosite-like mineral, probably derived from the alteration of pyrite, are also scattered throughout the unit along bedding planes or joints. The jarosite-like mineral, which contains small amounts of sodium, is referred to as jarosite elsewhere in this paper. The mudstone and carbonaceous shale that generally underlie the K/T boundary claystone are probably of fluvial and pond origin, as indicated by their laminated character and content of macerated plant debris. The sediments must have been deposited along low-relief flood plains in an exceedingly low-energy regime, as indicated by the small amount and small size (<100 μm) of included detrital mineral grains.

At three sites, one in New Mexico (Sugarite) and two in Colorado (Carmel and Madrid East), the rocks below the K/T boundary consist of either coal or impure coal. At another site about 12 km south of Trinidad, Colorado (Clear Creek South), the rock unit below the boundary claystone is very fine grained sandstone rather than the typical carbonaceous shale, mudstone, or impure coal.

The mineralogic composition of the hydrofluoric-acid insoluble residues from 11 samples of mudstone or impure coal from below the K/T boundary claystone in the Raton basin are listed on Table 1. The proportions of quartz, metaquartzite, quartzite, chert, feldspar, and nonopaque heavy minerals were determined in each sample by point-counter analysis (Chayes, 1956) of 1,000 grains in oil-immersion mounts. Quartz grains constitute the chief detrital component of the mudstone (99 to 77 percent); metaquartzite-quartzite, chert, and feldspar occur in minor amounts. Zircon, tourmaline, epidote, hornblende, and garnet, occur only in trace amounts. Authigenic minerals include rutile and clusters of pink quartz crystals, and perfectly formed colorless quartz crystals were found in samples of carbonaceous shale from 0 to 2 cm below the boundary claystone at the Clear Creek North site in Colorado.

Interestingly, a few shock-metamorphosed mineral grains were identified in the acid-insoluble residue obtained from carbonaceous shale 0 to 2 cm thick that directly underlies the boundary claystone bed at the Clear Creek North, Starkville South, and Madrid East sites in Colorado (Fig. 2). The rare shocked quartz grains probably were introduced into the carbonaceous shale either by gravity along roots of plants or by other bioturbation processes. The rare shocked quartz grains were found by splitting pieces, a few millimeters thick, of the carbonaceous shale from the base of the boundary claystone directly into a new plastic beaker. New plastic beakers and nylon bolting cloth (for screening) were used to eliminate the possibility of intralaboratory contamination. The shale was disaggregated and dissolved by ultrasonic scrubbing in a 5 percent solution of hydrofluoric acid.

At other K/T sites in the Raton basin listed on Table 1, the mudstone and impure coal underlying the K/T boundary claystone are essentially free of shock-metamorphosed minerals. No shocked quartz grains, as defined by multiple sets of shock lamellae, were seen during a point-counter analysis of 2,000 grains from a special set of samples collected from rocks directly below the K/T boundary at numerous sites.

Similarly, grains of shock-metamorphosed quartz have not been identified in rocks directly below the K/T boundary claystone at K/T boundary sites in Wyoming. No shocked grains were found during study of 2,000 grains in samples of mudstone from 1 to 9 cm below the boundary claystone at the Teapot Dome and Dogie Creek sites. Point-counter analyses of grains from each of these sites are shown in Table 1. Grains of quartz made up about 90 percent of the mineral assemblage, and chert, metaquartzite-quartzite, and feldspar constituted the remainder.

Two quartz grains, each having two poorly defined sets of planar lamellae, were found in the insoluble residue of mudstone from 16 to 20 cm below the boundary claystone at the Brownie Butte, Montana, K/T boundary site. These grains were reported (Izett and Pillmore, 1985a) as being shocked, but it should be emphasized that the microstructures were poorly developed and could just as well be of tectonic origin. I restudied splits of the samples from several stratigraphic levels below the boundary claystone at this site, but did not see any quartz grains with well-developed multiple sets of planar lamellae during a point-counter analysis of 2,000 grains. Point-counter analyses of 1,000 mineral grains from each sample and from as many as three stratigraphic levels below the K/T boundary are listed in Table 1.

The K/T boundary in continental sedimentary rocks of the Raton basin was placed at the abrupt stratigraphic disappearance of several palynomorphs (Tschudy and Tschudy, 1986). Near Trinidad, Colorado, in the Raton basin, the paleontologic K/T boundary lies about 20 m above the base of the Raton Formation

TABLE 1. RELATIVE ABUNDANCE OF MINERAL TYPES IN SEDIMENTARY ROCKS JUST BELOW THE K/T BOUNDARY CLAYSTONE IN THE WESTERN UNITED STATES*

Locality Rock Type (Interval)	QTZ (%)	CHT (%)	M-Q (%)	AF (%)	PF (%)	HB (%)	ZR (%)	EP (%)	AL (%)	TO (%)	GA (%)	RU (%)
Berwind Canyon (Colo.)												
Mudstone (0-2 cm)	87	2	5	5	Tr	Tr	Tr	0	0	Tr	Tr	0
Mudstone (13-15 cm)	80	3	7	9	Tr	0	Tr	0	0	0	Tr	0
Clear Creek North (Colo.)												
Carb. shale (0-2 cm)	91	1	6	1	Tr	0	Tr	0	0	Tr	0	0
Mudstone (1.5 m)	67	10	8	14	Tr	0	0	Tr	0	0	Tr	0
Starkville South (Colo.)												
Carb. shale (0-2 cm)	92	1	6	0	0	0	T	0	0	0	0	0
Mudstone (60 cm)	90	4	3	2	Tr	0	0	0	0	0	0	0
Long Canyon (Colo.)												
Impure coal (0-2 cm)	99	0	Tr	0	0	0	Tr	0	0	0	0	0
Madrid East (Colo.)												
Impure coal (0-2 cm)	98	0	1	0	0	0	Tr	0	0	0	0	Tr
Raton (N. Mex.)												
Mudstone (0-2 cm)	87	6	4	1	1	Tr	Tr	0	0	0	Tr	Tr
Mudstone (12-13 cm)	85	2	3	9	Tr	0	Tr	Tr	0	0	0	0
Climax Canyon (N. Mex.)												
Mudstone (0-5 cm)	77	4	7	10	1	0	Tr	Tr	0	0	Tr	0
Teapot Dome (Wyo.)												
Mudstone (0-2 cm)	90	3	2	3	Tr	Tr	Tr	0	0	0	0	0
Mudstone (23 cm)	90	2	4	3	Tr	0	Tr	Tr	0	0	0	0
Dogie Creek (Wyo.)												
Mudstone (0-2 cm)	91	1	2	4	1	Tr	Tr	Tr	0	0	0	0
Brownie Butte (Mont.)												
Mudstone (5-7 cm)	57	10	20	9	3	Tr	0	Tr	Tr	0	0	0
Mudstone (7-10 cm)	78	10	11	Tr	0	Tr	0	0	Tr	0	0	0
Mudstone (16-20 cm)	76	11	12	Tr	Tr	0	0	0	0	0	0	0

*Relative abundances based on a point-counter analysis of 1,000 mineral grains from each sample. Distance below K/T boundary claystone shown in parentheses. QTZ, quartz; CHT, chert; M-Q, metaquartzite and quartzite; AF, potassium feldspar; PF, plagioclase feldspar; HB, hornblende; ZR, zircon; EP, epidote; AL, allanite; TO, tourmaline; GA, garnet; RU, rutile; Tr, trace; Figures 2 and 45 show location of sites.

(Pillmore and Tschudy, 1983). On the basis of the regional extinction of several palynomorph taxa, including *Proteacidites,* the boundary was placed by Tschudy and others (1984) and Tschudy and Tschudy (1986) at the top of the boundary claystone. This was an arbitrary choice because only rare palynomorphs have been recovered from the boundary claystone, and these might have been reworked into the unit from underlying Cretaceous sediments. The rocks just above the boundary claystone, which contain abundant sapropel and decaying vegetation, were termed the "barren-and-recovery interval" by Tschudy and Tschudy (1986, p. 668). An abrupt increase in the ratio of fern spores to angiosperm pollen also occurs at this horizon. The abrupt increase in fern spores above the boundary claystone in the Raton basin has been called the "fern spore spike." In contrast, Sweet (1988) reported that angiosperm pollen are dominant in rocks just above the K/T boundary claystone in Alberta, Canada.

The supposition that the pollen-defined K/T boundary in continental rocks of the Western Interior is coeval with the fossil-defined K/T boundary in marine rocks is based on two observations: (1) both boundaries occur in a reverse polarity magnet-

ozone, 29R (Shoemaker and others, 1987, p. 148), and (2) anomalously large amounts of platinum-group metals and a spike of shocked minerals (Izett, 1987a, 1988) are restricted essentially to a thin unit (the K/T impact layer) at both marine and continental sites.

The disappearance of the dinosaurs has been widely used in western North America to define the K/T boundary, but dinosaur remains have not been found in Upper Cretaceous rocks in the Raton basin. Curiously, dinosaur remains do occur in Upper Cretaceous continental rocks just to the south and north of the Raton basin in the San Juan and Denver basins.

K/T BOUNDARY CLAYSTONE BED

The most distinctive microstratigraphic unit in the K/T boundary interval in the Raton basin of Colorado and New Mexico was discovered by C. L. Pillmore of the U.S. Geological Survey and called by him the " K/T boundary clay." Although Pillmore and his colleagues (1984) referred to the bed lithologically as clay, it is indurated and should be termed a "claystone." Within the basin, its stratigraphic level was established by Pillmore, and the claystone has been found at about 20 different sites. These sites occur from along North Ponil Creek, New Mexico, to near Ludlow, Colorado, a distance of about 80 km (Fig. 2). The locations of many of these sites were shown on a map presented by Pillmore and Flores (1987).

The K/T boundary claystone bed is generally a thin layer of macroscopically uniform claystone that weathers to light shades of gray and brown. The overall light color of the claystone stands in marked contrast to subjacent dark gray carbonaceous shale, mudstone, and coal, and to the superjacent coal (Fig. 1). The claystone is moderately indurated and, when crushed, breaks into angular fragments and crumbs, as do flint clay and semiflint clay. The claystone bed is generally 1 to 2 cm thick, and for such a thin unit, it is astonishingly uniform in regional thickness. At a few sites, including one along Long Canyon 10 km southwest of Trinidad, Colorado, the bed is abnormally thick, about 3.0 cm.

Because the boundary claystone forms a thin, discrete unit within a coal bed, or directly below a coal bed, it is similar to kaolinitic claystone partings in coal sequences called "tonsteins" by European geologists (Williamson, 1970). Tonsteins are thought by some geologists (Bohor and Pillmore, 1976), in some instances without convincing proof, to be altered volcanic ash beds. The boundary claystone is also similar in its stratigraphic setting and certain lithic characteristics to claystone layers below coal beds termed "underclays" and "flint clays" (Huddle and Patterson, 1961; Smyth, 1984).

The clay minerals composing the boundary claystone were studied and identified by x-ray diffraction methods by Pollastro and Pillmore (1987) and, in this study, by C. G. Whitney (U.S. Geological Survey, written communication, 1986), R. A. Sheppard, and the writer. Five samples from the Raton basin and one from the Brownie Butte, Montana, site were analyzed by Whitney. The analyses show that the claystone consists of well-crystallized kaolinite and a small amount of random and ordered I/S mixed-layer clay. The amount of I/S mixed-layer clay relative to kaolinite in the boundary claystone is considerably greater at K/T boundary sites in Wyoming, Montana, and Canada as compared to sites in the Raton basin. The greater amount of I/S mixed-layer clay at sites in Wyoming, Montana, and Canada causes the claystone to swell or expand when hydrated and to crack when dried. Thus, the boundary claystone at Wyoming, Montana, and Canada sites is similar, in these respects, to swelling volcanic bentonites.

In southwestern Saskatchewan, Canada, at the Frenchman River site of Lerbekmo and others (1987), the boundary claystone consists mainly of smectite and small amounts of kaolinite and a zeolite mineral having a heulandite structure, probably clinoptilite. The zeolite was identified by R. A. Sheppard and the writer on an x-ray diffractogram of the claystone. The occurrence of the zeolite in the boundary claystone at the Frenchman River site suggests that the geochemical environment in which the claystone formed was more alkaline than that at the Morgan Creek site and at sites in the Raton basin. In contrast, the boundary claystone at the Morgan Creek site of Nichols and others (1986), also in southwestern Saskatchewan, consists chiefly of kaolinite.

The texture of the claystone, as viewed with a biologic-type stereomicroscope, is extremely fine grained. Typically the claystone contains carbonaceous material, and it is similar in this respect to tonsteins. At most sites, the boundary claystone contains centimeter-long horizontal vitrinite laminae and convoluted swirls of vitrinite that may be deformed roots and rootlets of plants (Smyth, 1984, Plate 37; Retallack, 1988). Scattered through the claystone are flecks of decomposed plant material so altered that their original nature barely can be recognized.

The claystone at most sites in the Raton basin contains scattered irregularly shaped inclusions (or clasts?) several hundred micrometers in length that are composed of a texturally and compositionally younger clay-rich material, which differs markedly from the host kaolinite (Fig. 3). The inclusions may be composed of a mixture of clay minerals including I/S mixed-layer clay and hydrous iron oxide. The clay-rich material that forms the inclusions contains significant amounts of iron and small amounts of potassium and calcium as determined with the energy dispersive system (EDS) of a scanning electron microscope (SEM). The inclusions originally may have been vugs formed from the complete decay of fragmental plant material. These vugs or voids in the progenitor of the claystone subsequently were filled with the clay and iron oxides. Fastovsky and McSweeney (1987, Fig. 2) illustrated vugs in Cretaceous and Tertiary boundary paleosols that are similar to those in the boundary claystone. In addition, Smyth (1984) described pedogenic vugs or pores in a kaolinitic claystone, the Olive Hill Fire Clay of Pennsylvanian age, that also resemble those in the boundary claystone. Fastovsky and others (1989) identified features in the K/T boundary claystone at Brownie Butte, Montana, as "coatings" and "hypocoatings." These features typically form from the translocation of clay minerals or iron oxides in a soil-forming environment.

The pervasive groundmass texture of the claystone, as viewed with a petrographic microscope, is cryptocrystalline; it differs in this respect from other studied tonsteins in the Raton basin, which are coarse-grained kaolinite. The texture is a micrometer-size aggregate of kaolinite granules and rare lenticules of complexly arranged accordian-like kaolinite booklets. The term "cryptotonstein" was applied by Retallack (1976, p. 390) to similar fine-grained kaolinitic claystone beds (tonsteins), and he suggested they possibly form by biochemical processes in soils. The claystone contains scattered patches, a few tenths of a millimeter in length, of a feathery-appearing clay mineral having a higher birefringence than the adjacent kaolinite groundmass. These are doubtless the clay inclusions seen under low magnification with a stereomicroscope and an SEM (Fig. 3).

Study of the material that composes the boundary claystone in the Raton basin with an SEM revealed that it mainly consists of a polygonal framework of kaolinite filled with micrometer-size kaolinite microspherules (Fig. 4A). The construction of the kaolinite framework is similar to the hexagonal close packing lattice. The texture of this honeycomb-like groundmass suggested to Pollastro and Pillmore (1987) that the precursor of the claystone was glass. But the microspherulitic texture also occurs in claystone that fills cells of degraded plant material and in cores of kaolinite spherules (Izett, 1987a). This evidence indicates that the texture probably did not result from the in situ alteration of glass, but instead resulted from the crystallization of a silica-alumina gel.

The texture of the boundary claystone, as seen with an SEM, from sites in Canada is not always microspherulitic as it is in the Raton basin. At the Morgan Creek site of Nichols and others (1986) in southwestern Saskatchewan, the claystone is composed of kaolinite, and its texture is uniformly microspherulitic. In contrast, the composition of the claystone at the Frenchman River site of Lerbekmo and others (1987) is smectite, kaolinite, and zeolite, and its texture reflects this mixed composition. The texture consists of a wavy open framework of smectite interspersed with microspherules of kaolinite.

Another texture seen within the boundary claystone in the Raton basin consists of a diffuse layer, several millimeters thick, of a compacted mass of kaolinite spherules and curved plate-like vermicules (Fig. 4B). In some specimens, the texture occurs in a layer in the middle of the boundary claystone. Single and groups of uncrushed kaolinite spherules occur near highly crushed masses, and fragments of kaolinite spherule walls are embedded in the microspherulitic groundmass. The kaolinite spherule walls and curved plates of crushed spherules, which form an essential part of the texture, typically contain a concentration along their convex sides of light brown submicrometer-size inclusions. The inclusions are rich in titanium as demonstrated using the EDS analyzer of an SEM.

The texture of the boundary claystone described in the foregoing paragraph was referred to as "imbricate" by Pollastro and Pillmore (1987) in reference to its superficial similarity to the compacted fabric of welded tuffs. The texture obviously formed by compaction, and in this respect it is similar to welded tuffs. However, the compaction of the boundary claystone occurred at low temperature, and the compaction of welded tuffs at high temperature. The overall texture of the claystone as described above suggests that it developed during repeated episodes of spherule growth and compaction, possibly in a soil-forming environment.

At most sites in the Raton basin, the boundary claystone is virtually free of detrital quartz and other mineral grains and consists of macroscopically uniform material. For example, study of thin sections and polished surfaces of the boundary claystone showed that no visible quartz grains occur in several hundred square centimeters of the claystone. At the Clear Creek North K/T boundary site, Colorado, only two quartz grains, not obviously associated with carbonaceous paleoroots (described in a following paragraph), were seen during study of 110 cm^2 of the boundary claystone. No quartz grains were seen in study of 100 cm^2 of the boundary claystone from the Berwind Canyon and Madrid East sites in Colorado (Fig. 2), except for a few grains that occur just below the contact with the overlying impact layer. Thus, the evidence shows that quartz and other mineral grains are either absent or exceptionally rare in many samples of the boundary claystone.

At one K/T boundary site in the Raton basin (Clear Creek North), the claystone appears macroscopically to be homogeneous and free of clastic mineral grains, but microscopic examination shows that it contains rare clastic quartz grains, some of which are shock-metamorphosed because they display multiple intersecting sets of planar lamellae. The quartz grains are concentrated along subvertical carbonaceous sedimentary structures (described in a subsequent paragraph). In some instances, the grains appear where clastic mineral grains in the overlying impact layer are most numerous. The presence of the quartz grains was revealed by polishing ground surfaces of the claystone with 6-μm diamond paste rubbed on the surface of the claystone with a soft paper tissue. The quartz grains stand in conspicuous relief above the softer surrounding claystone owing to their greater hardness. Grains as small as a few tens of micrometers may be revealed under the microscope using this technique.

At several places, such as the Clear Creek South site in Colorado, the boundary claystone is composed of sandy claystone. This fact clearly indicates it is composed of water-deposited material and not primary air-fall impact ejecta as some believe (Pillmore and others, 1984). At the Clear Creek South site, the boundary claystone rests on a white, fine-grained sandstone, and the contact between the units is gradational through an interval of a few millimeters. At the south end of the outcrop, the claystone is as much as 5.0 cm thick and contains about 20 percent quartz grains. The sand grains are set in a wavy-bedded, clayey matrix composed mainly of kaolinite and I/S mixed-layer clay. Near the middle of the outcrop, the boundary claystone is about 2.0 cm thick and consists of two distinct layers, each of which formed during a separate depositional stage (Fig. 5).

The lower 1.5 cm of claystone, which formed during stage 1, contains scattered quartz grains set in the claystone matrix.

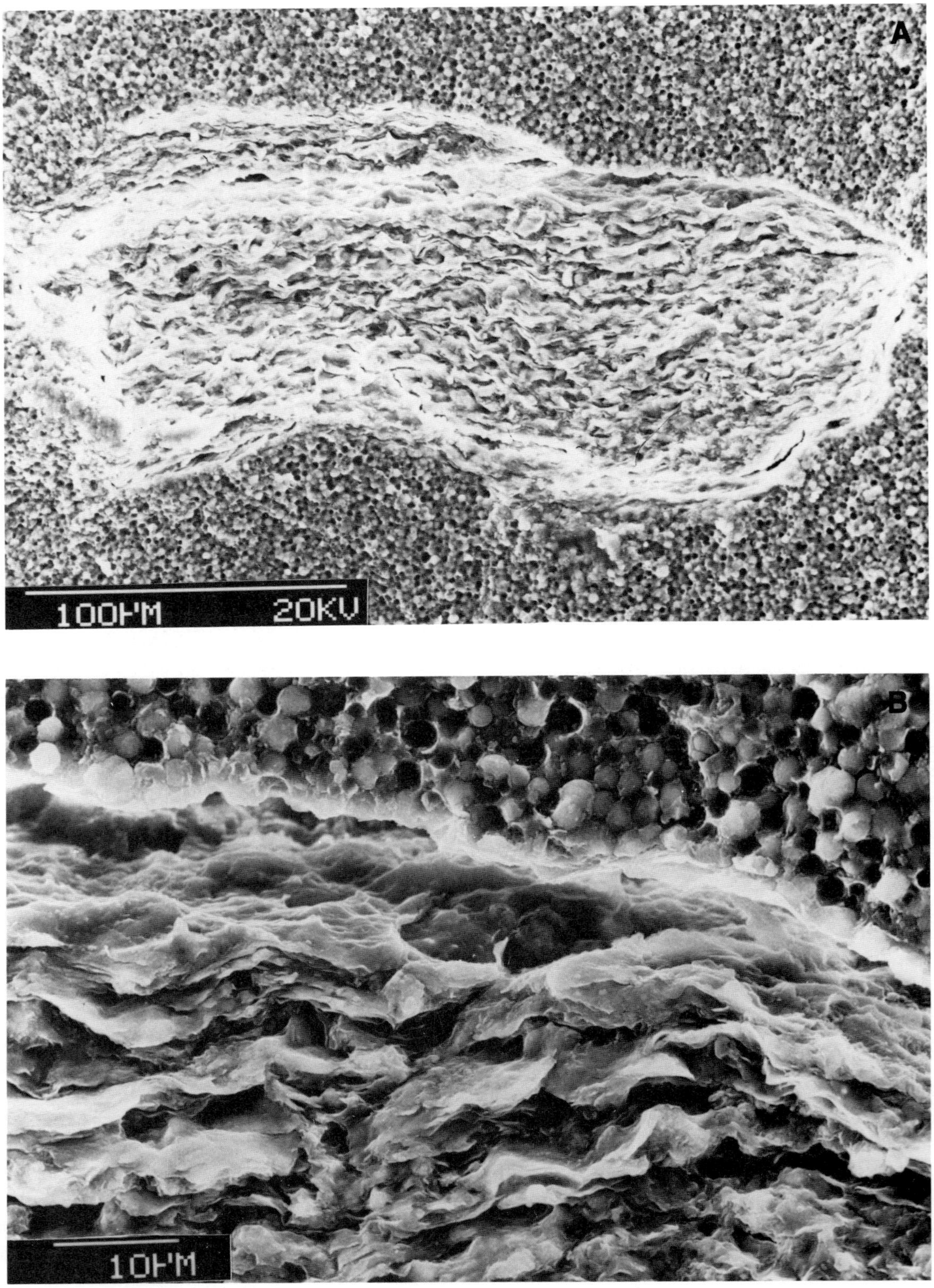

Figure 3. Scanning electron microscope (SEM) photomicrographs showing claystone inclusion in the K/T boundary claystone from the Raton, New Mexico, site. A, low-magnification view of inclusion about 0.3 mm in length that may have formed by the infilling of a void with I/S mixed-layer clay and iron oxide. Some of these voids were incompletely filled during diagenesis. B, high-magnification view of contact between inclusion of 3A. Note microspherulitic texture of K/T boundary kaolinitic claystone and the flaky texture of the I/S mixed-layer clay of the inclusion.

Figure 4. Microscopic texture of the K/T boundary claystone from the Starkville South K/T boundary site, about 8 km south of Trinidad, Colorado. A, texture as revealed by the SEM is a microspherulitic polygonal framework of kaolinite filled with kaolinite microspherules. B, texture as seen optically with incident light is locally a compacted mass of crushed and uncrushed kaolinite spherules, some of which appear hollow. Arrow points to a clastic fragment of claystone that seemingly is rimmed by another generation of clay.

Figure 5. Polished slab of the K/T boundary interval from the Clear Creek South K/T boundary site, about 12 km south of Trinidad, Colorado. The boundary claystone bed, impact layer, and boundary coal bed comprise the boundary interval. The boundary claystone consists of two subunits of reworked claystone that typically contain planar shreds of vitrinite and a considerable amount of unshocked quartz and feldspar grains. Rare quartz grains show multiple sets of shock lamellae. Two large angular clastic claystone fragments composed of unreworked boundary claystone shown at arrows. One clastic claystone fragment is truncated by the overlying unit.

Among the thousands of such grains studied in oil-immersion mounts were two that exhibited multiple sets of shock lamellae. The upper 0.5 cm of claystone deposited during stage 2 contains more disseminated planar shreds of vitrinite than the claystone deposited during stage 1, as shown by its dark gray color (Fig. 5). Study of oil-immersion mounts of clastic grains separated from the claystone deposited during stage 2 shows that shock-metamorphosed minerals are about five times as abundant as in the underlying claystone deposited during stage 1. In contrast, the impact layer that overlies these units of sandy claystone locally contains a normal concentration of shocked quartz grains (as much as 1 percent).

Lithologic evidence that the boundary claystone in the Raton basin, which locally consists of water-transported (not primary air-fall) material, is not confined to the Clear Creek South site. At the Sugarite and Sugarite South sites in New Mexico and the Carmel site in Colorado, the boundary claystone also consists of water-transported material, but because the subjacent rocks are coal or impure coal, the claystone contains only rare quartz and feldspar grains. Where the boundary claystone consists of sandy claystone, it typically and pervasively contains planar shreds of vitrinite preferentially oriented parallel to bedding. The abundance of shreds and flakes of vitrinite in K/T boundary interval rocks is considered in this paper as evidence that the sediments consist of reworked material.

Evidence that the boundary claystone consists of water-transported reworked material is afforded by its texture, which is fragmental, and it occurs in two forms. The first form is obvious and can be seen without the aid of a microscope. At the Clear Creek South site in Colorado, angular claystone fragments (Fig. 5) as large as 1.0 cm occur in the claystone. They have a texture identical to that of the average (unreworked) boundary claystone. Some of these fragments are truncated by an overlying erosion surface at the base of the overlying (stage 2) unit. Thus, the evidence indicates that the claystone fragments probably were derived by erosion of boundary claystone that had sufficient strength to survive gentle erosion and transportation. The amount of time between the formation of the claystone and its subsequent erosion and transportation is unknown, but it may have been brief. These relations imply that the boundary claystone was deposited, partially lithified, and eroded prior to the arrival of shock-metamorphic minerals in the impact layer.

The second form of fragmental texture can be seen only in thin sections of the claystone. Angular fragments of cryptocrystalline kaolinite, nearly as large as 1.0 mm, generally occur in the lower part of the claystone (Fig. 4B). Some fragments have not been moved far because their margins can be matched with adjacent fragments. This fragmental texture is restricted to the boundary claystone and does not occur in the overlying impact layer, suggesting that the process responsible for the fragmentation of the boundary claystone occurred prior to the deposition of the impact layer.

The fragmental nature of the boundary claystone is particularly well developed at two sites in Wyoming. One site is near Teapot Dome in the NW¼,Sec.23,T.38N.,R.77W., Converse County (Wolfe and Izett, 1987). The second site is along Dogie Creek in the SW¼SE¼,Sec.5,T.38N.,R.65W., Niobrara County (Bohor and others, 1987). At the Teapot Dome K/T boundary site, centimeter-size clastic pellets composed of kaolinitic claystone occur along a reddish brown, goyazite-mineralized layer at the top of the boundary claystone. Goyazite is a hydrous, strontium-rich alumino-phosphate mineral similar to crandallite, which is the calcium-rich analogue.

Additional evidence that the K/T boundary claystone consists of water-transported material occurs at the Wyoming sites. At the Dogie Creek site, the boundary claystone locally grades downward into white, very fine grained sandstone. The lower part of the boundary claystone contains an appreciable amount (several percent) of quartz grains. No shock-metamorphosed quartz grains were seen in the insoluble residue of this claystone during a point-counter analysis of 2,000 grains.

The claystone of the pellets is composed of kaolinite that is much harder than that of the normal boundary claystone. It is similar to flint clay in its hard and brittle nature. Furthermore, the claystone pellets show pedogenic slickensides (Gray and Nicklesen, 1989) on some surfaces as does flint clay. The microscopic texture of the claystone of the pellets is formed of randomly oriented flakes of kaolinite, unlike the microspherulitic kaolinite of the boundary claystone. The shape of the kaolinite pellets in the goyazite-mineralized layer in sections cut normal to bedding is ovoidal to lenticular. As seen on bedding planes, the pellets are approximately equidimensional. Some claystone pellets contain goyazite spherules (Fig. 6) that generally exist near the margins of the kaolinite pellets. A few pellets are penetrated by veins of goyazite, and some contain irregular-shaped goyazite cavity fillings, carbonaceous plant material, and quartz grains.

The claystone of one pellet, about 1.0 mm in diameter, was dissolved in dilute hydrofluoric acid (~5 percent). The hydrofluoric acid insoluble residue consisted of about 95 percent quartz, a few percent feldspar, chert, and quartzite, and rare zircon, tourmaline, and green-brown hornblende. No shocked quartz grains, as defined by multiple sets of shock lamellae, were seen during a point-counter analysis of 2,000 grains in an oil-immersion mount.

Bohor (1988a, 1988b) speculated that the kaolinite pellets were originally glass lapilli or bombs of impact origin. If the precursor material of the pellets were glass, as Bohor imagined, the clay alteration products of the glass should have, if Pollastro and Pillmore (1987) were correct, a microspherulitic texture. As described in previous paragraphs, the texture of the claystone pellets is not microspherulitic. Additional evidence conflicting with the idea that the pellets were originally impact glass lapilli is that the pellets were flattened during an early stage of soft-sediment diagenesis. If the pellets were originally impact glass, they surely would not have altered to clay early in their diagenetic history, and thus, they probably would have retained their original shapes (not flattened lens-shaped objects). Under ambient conditions, impact glass almost certainly would be stable for a considerable time following deposition. Be this as it may, the most important evidence refuting Bohor's speculation that the precursor of the kaolinite pellets was glass is that some of the pellets contain carbonaceous fossil plant material.

Locally, the boundary claystone contains peculiar carbonaceous structures that can be seen on sawed and polished slabs of the claystone cut normal to bedding (Figs. 1 and 7). These structures are rare and best developed at the Clear Creek North site. They are subvertical to vertical and consist of poorly defined, disjointed, and deformed (accordion-like) trains of vitrinite. Some of them are funnel-shaped in cross section; that is, they appear to flare out at the contact with the overlying impact layer (Fig. 7B). Where they flare out at the contact with the overlying impact layer, carbonaceous material may be concentrated. Others are smeared out horizontally at their upper ends within the claystone (Fig. 7A). Some of these structures extend from the base to the top of the boundary claystone bed. However, the carbonaceous structures have not been seen to cut the fine claystone or vitrinite laminations of the overlying impact layer. In sections cut parallel to bedding, most of the structures are irregular in plan, but a few seem to be roughly circular. Quartz and other clastic mineral grains are generally restricted to the subvertical structures and have not been seen in the horizontal carbonaceous laminae.

The origin of the carbonaceous structures is equivocal, but they probably are the remains of roots and rootlets of plants that grew on a surface at the top of the boundary claystone prior to deposition of the overlying impact layer. Following the depositional event that formed the boundary claystone, quartz grains and clay filled the voids in decomposed roots of plants.

In addition to the carbonaceous root-like structures just described, the boundary claystone contains large fragments of fossil plant material. At several places, it contains millimeter-size rounded fragments of cellular fusinite that generally occur in the lower part of the bed. At the Madrid East site near Madrid, Colorado, a centimeter-size chunk of rounded, carbonized, cellular plant material occurs in the upper part of the claystone bed (Fig. 8). The claystone matrix that surrounds the chunk of fossil plant material is totally free of detrital quartz and feldspar grains.

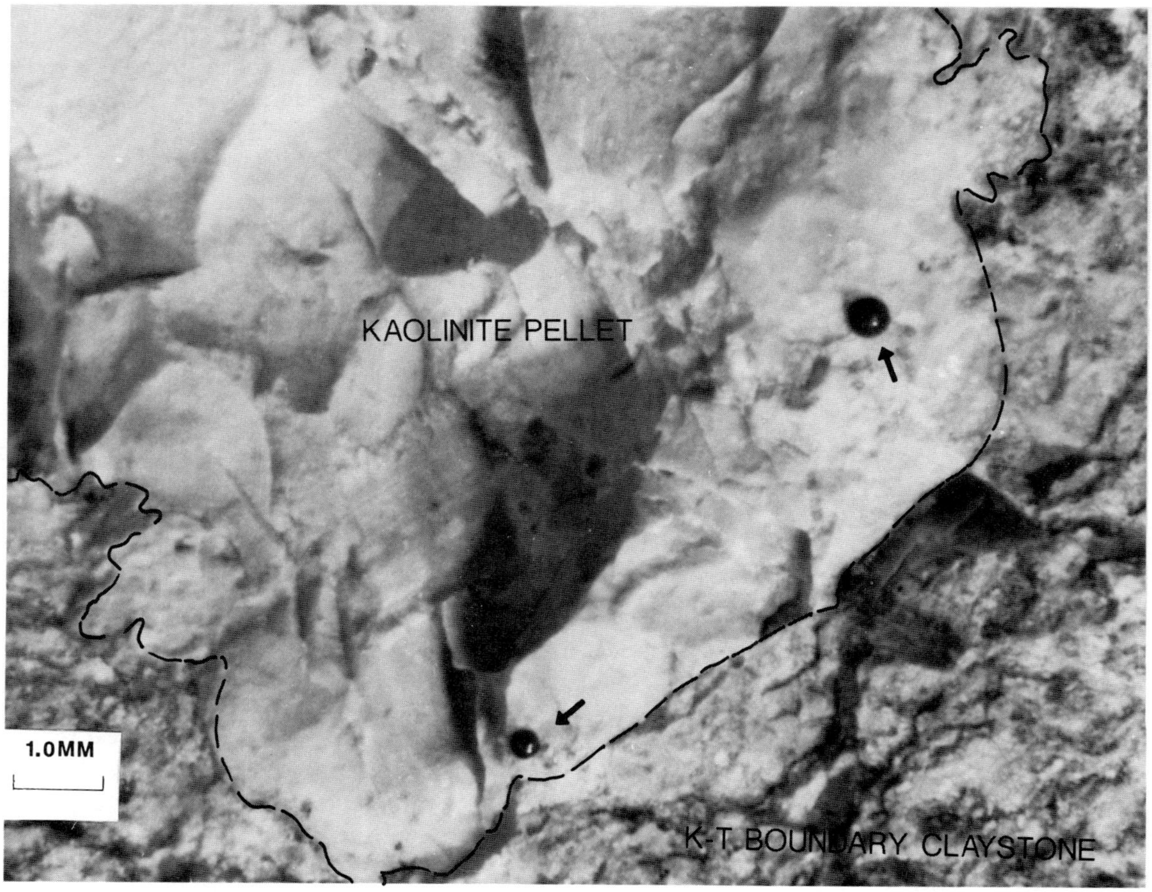

Figure 6. Two goyazite spherules (shown at arrows) in a kaolinite pellet in the K/T boundary claystone from the Teapot Dome, Wyoming, K/T boundary site of Wolfe and Izett (1987). Boundary between pellet and surrounding kaolinite of the boundary claystone shown by dashed line. Large goyazite spherule about 0.5 mm in diameter.

The boundary claystone also contains fossil plant material at Wyoming and Montana K/T boundary sites. At the Teapot Dome site in Wyoming, the claystone contains considerable amounts of goyazite-replaced plant material. Among the goyazite-replaced plant fibers are goyazite mammillary masses and microspherules only a few tens of micrometers in diameter (Fig. 9). In addition, I found fossil roots and rootlets along bedding planes and the corm (nodule on a root) of a plant, possibly a sedge or water lily (J. A. Wolfe, written communication, 1987). Also at this site, the base of the boundary claystone commonly contains abundant flecks and fragments of detrital carbonaceous plant material. A coalified twig 1.5 cm long also was found in the claystone at the Teapot Dome site. At the Brownie Butte, Montana, site, the claystone contains fossil impressions of twigs as long as 2 cm.

In samples of the boundary claystone from the Morgan Creek site in Saskatchewan, Canada (Nichols and others, 1986), I found blebs of clear, yellow amber as much as 2.5 mm in diameter. The blebs of amber occur at various stratigraphic positions in the claystone, and presumably formed from drops of conifer sap that sporadically fell directly into the accumulating sediment that now forms the boundary claystone. The occurrence of such blebs of amber in the boundary claystone is of considerable interest because Berner and Landis (1987) determined that the composition of gas in bubble inclusions in Eocene and Oligocene amber is similar to that of present-day air. If the blebs of amber contain bubble inclusions and entrapped gas, the composition of this gas may shed light on the composition of the Cretaceous/Tertiary boundary atmosphere.

Microstratigraphic relations

At most places in the Raton basin, the K/T boundary interval consists of the normal sequence of microstratigraphic units, which includes, in ascending order, carbonaceous shale or coal, the boundary claystone, the impact layer, and the boundary coal bed. However, at three Colorado K/T boundary sites, Clear Creek North, Madrid Railroad, and Carmel (Fig. 2), the boundary claystone pinches out into carbonaceous shale. This stratigraphic relation is particularly well exposed at the Carmel site in a stream cutbank along Long Canyon, west of Trinidad, Colorado (Fig. 2). At this site, the impact layer and boundary

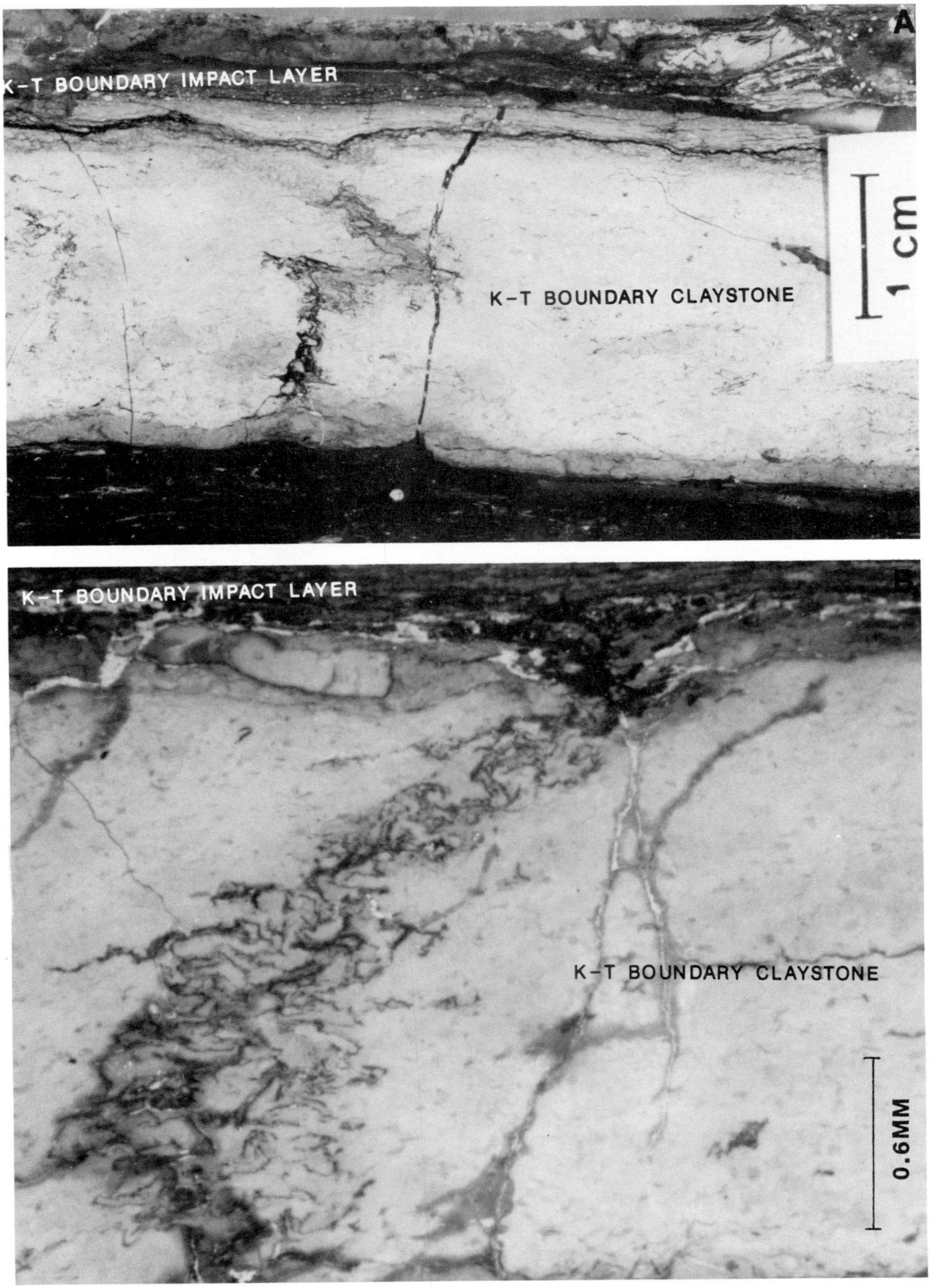

Figure 7. Carbonaceous rootlike structures in the K/T boundary claystone from the Clear Creek North K/T boundary site, 10 km south of Trinidad, Colorado. A, rootlike structure in boundary claystone smeared out near middle of the bed. B, rootlike structure contains a concentration of carbonaceous material at the contact between the boundary claystone and impact layer. Clastic quartz and other mineral grains occur along such features but not in the surrounding claystone.

Figure 8. Polished slab of the K/T boundary interval from the Madrid East K/T boundary site, near Madrid, Colorado. The boundary interval consists of carbonaceous shale, the boundary claystone, and the impact layer. The boundary coal was lost during preparation. A rounded, coalified wood fragment shown at arrow. Bedding in boundary claystone warped over the fragment.

coal bed can be traced for about 30 m intermittently along the outcrop. The impact layer can be recognized macroscopically owing to its usual content of tiny barley-shaped kaolinite pellets referred to by some as graupen. A generalized diagrammatic section prepared from outcrop observations and polished slabs of the K/T boundary interval illustrates the stratigraphic relations (Fig. 10). The diagram shows that the normal sequence of bqundary units can be traced along the outcrop to a point where the boundary claystone pinches out into carbonaceous shale. Further along the outcrop, the boundary claystone lithology occurs only as thin, wispy lenticules enclosed in carbonaceous shale, and the impact layer rests on carbonaceous shale free of lenses of boundary clay lithology. Where the boundary claystone pinches out, the impact layer is unusually thick, about 20 mm. The boundary coal bed, which is only about 1 cm thick, can be traced the length of the outcrop and provided an excellent reference horizon.

Contact relations

Megascopically, the contact between the boundary claystone and underlying rocks, chiefly dark gray carbonaceous shale, appears sharp, conformable, and undulatory (Fig. 1A). However, microscopic study of sawed, polished surfaces of boundary interval specimens reveals that the contact is actually gradational in detail. In many specimens, the homogeneous boundary claystone grades downward into carbonaceous shale or mudstone through an interval of about 5 mm. The lowermost part of the boundary claystone, as viewed on polished slabs cut normal to bedding, typically has a peculiar fluidal planar shear texture. The undulatory nature of the contact probably resulted from original surface irregularities and differential soft-sediment compaction and deformation during diagenesis of the carbonaceous-rich sediment.

The gradational character of the contact between carbonaceous shale and overlying boundary claystone is well displayed in samples from the Teapot Dome, Wyoming, site. At this site, the boundary claystone consists of light brown kaolinite, and it grades downward into medium gray claystone that contains abundant shreds and flecks of carbonaceous material and sparse quartz grains.

In the Raton basin, the contact between the boundary claystone and the overlying impact layer separates two distinctly different lithologies. The contact is here defined to be at (1) the lowest appearance of a concentration of shocked minerals, (2) the base of the lowest horizontal vitrinite laminae, or (3) the base of a billowy claystone unit. Generally these three stratigraphic keys occur at about the same horizon. The contact is usually sharp, but

Figure 9. Goyazite spherules (shown at arrows) among goyazite-replaced plant fibers in the K/T boundary claystone from the Teapot Dome, Wyoming, K/T boundary site of Wolfe and Izett (1987).

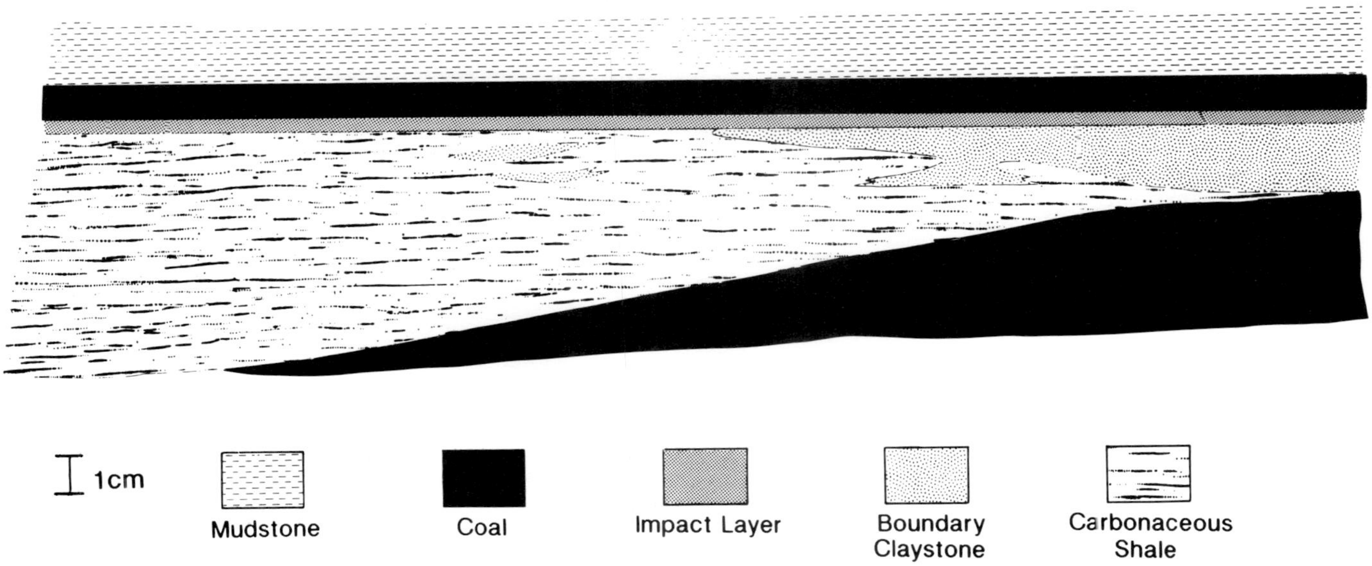

Figure 10. Diagrammatic section showing stratigraphic relationships in the K/T boundary interval at the Carmel K/T boundary site, southwest of Trinidad, Colorado. Where the boundary claystone pinches out, the impact layer rests directly on carbonaceous shale.

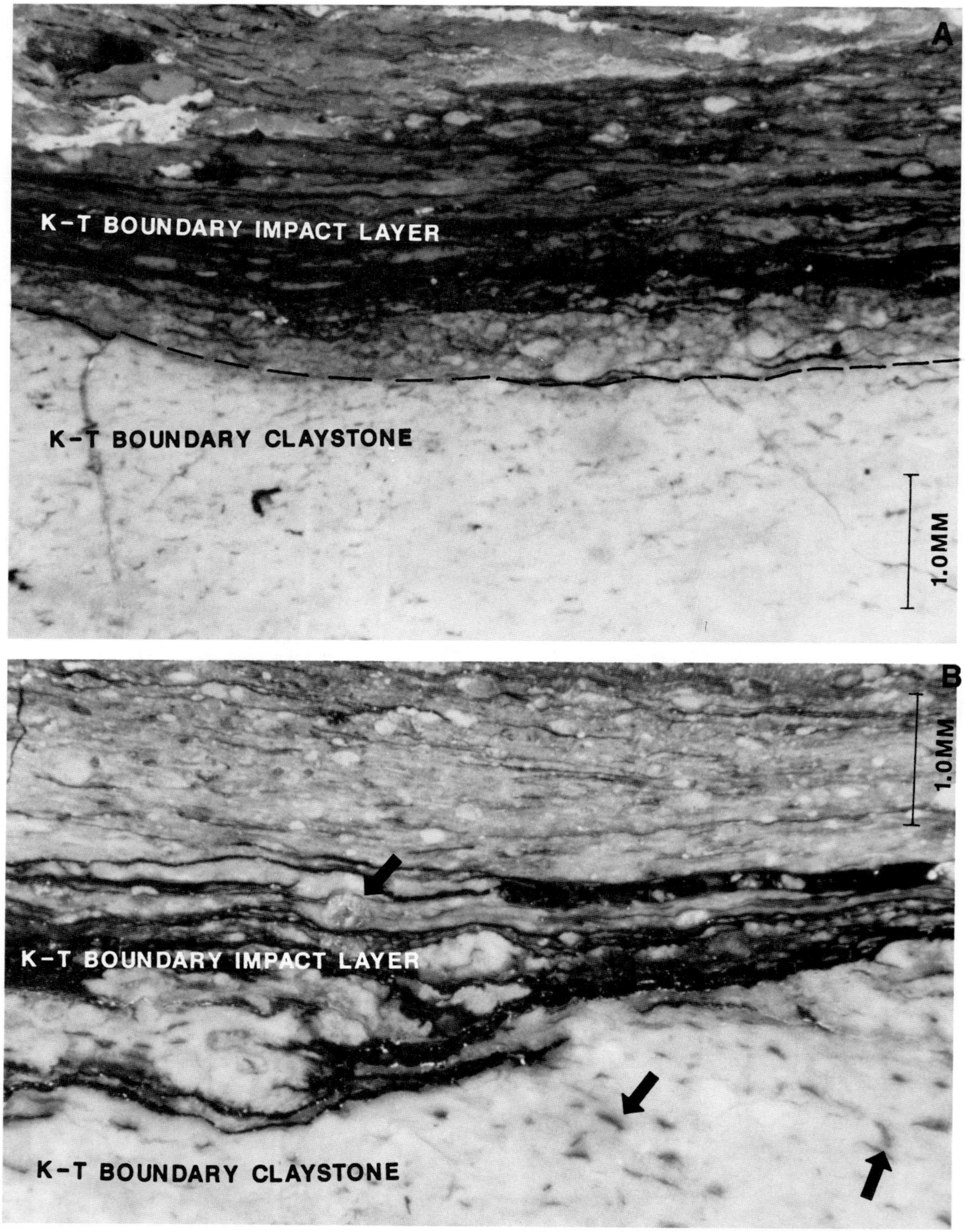

Figure 11. Contact between the K/T boundary claystone and impact layer. A, billowy nature of the lowermost part of the impact layer in specimen from the Clear Creek North K/T boundary site, a few kilometers south of Trinidad, Colorado. Dashed line shows contact. B, dark patches of I/S mixed-layer clay in boundary claystone and large shocked quartz grain (at arrows) in impact layer in specimen from the Berwind Canyon K/T boundary site, about 20 km northwest of Trinidad, Colorado.

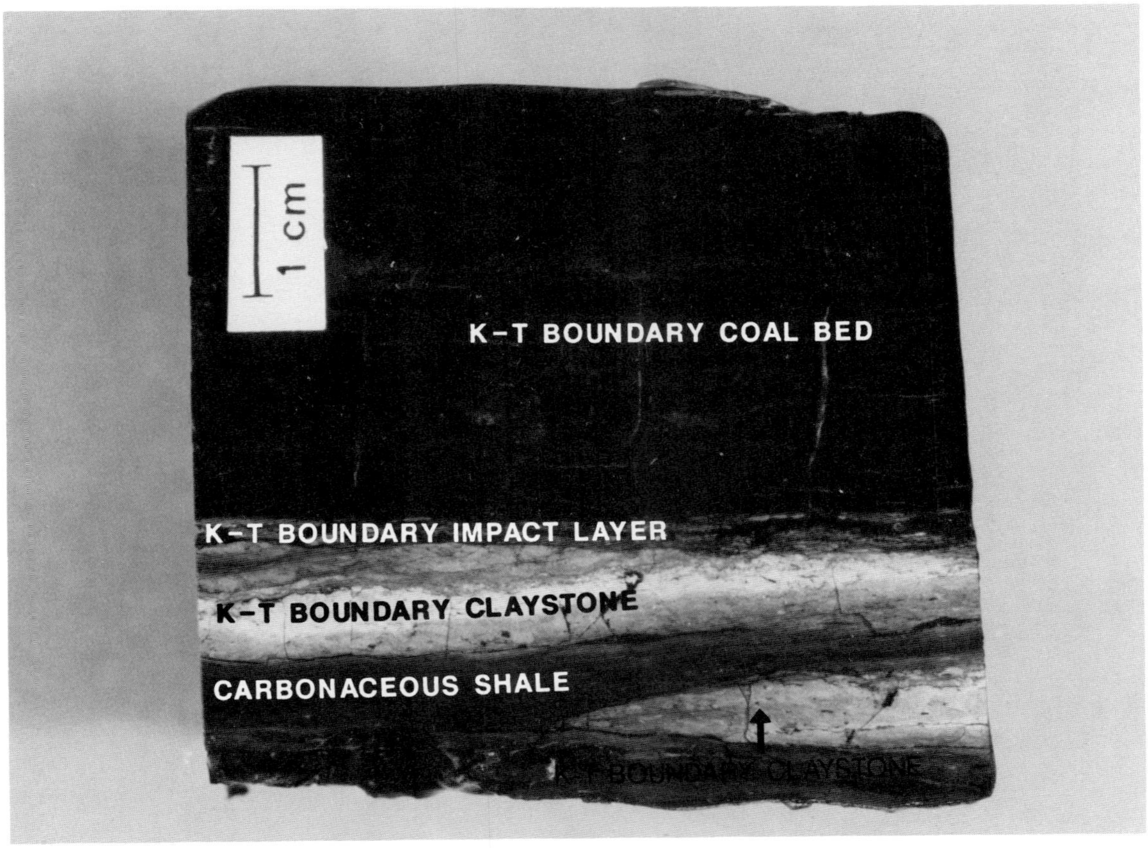

Figure 12. Polished surface of a slab of the K/T boundary interval from the Clear Creek North site, a few kilometers south of Trinidad, Colorado. Microstratigraphic units shown, in ascending order, are (1) carbonaceous shale containing a lens of boundary claystone lithology overlain by another layer of boundary claystone lithology, (2) the K/T boundary impact layer, and (3) the K/T boundary coal bed.

at some places it appears gradational, though only within a 1 to 2 mm interval. Within this interval the claystone appears billowy or turbated and seems to have been degraded into proto-pellets (Fig. 11). At most sites, the contact separates relatively structureless homogeneous claystone of the boundary claystone, which contains swirls of vitrinite, from overlying finely laminated claystone of the impact layer, which contains barley-shaped graupen and vitrinite laminae (Fig. 1). This observation implies that a preimpact deformational process locally contorted vitrinite-replaced plant material, mostly roots and rootlets, in the boundary claystone prior to the deposition of the impact layer.

Locally at the Clear Creek North site in Colorado, the impact layer rests directly on carbonaceous shale rather than the boundary claystone, and the contact between the units is a diastem or paraconformity (Fig. 12). The microstratigraphic relations are more complex than this because the carbonaceous shale contains lenses a few millimeters thick of boundary claystone lithology. These relations indicate that carbonaceous shale and boundary claystone may have been reworked prior to deposition of the impact layer.

At the Berwind Canyon and Starkville North sites in Colorado, the contact between the boundary claystone and impact layer is irregular on a small scale. At the Berwind Canyon site, impact-layer claystone fills both irregular and dike-like small-scale structures that project down into the boundary claystone (Fig. 13A). These structures, which are rare, are neither outlined by carbonaceous material nor are they root-like in configuration. The dike-like geometry in cross section suggests they may be cracks (desiccation?) that formed on top of the boundary claystone and were subsequently filled with impact-layer sediment. Rare clastic quartz and other mineral grains are scattered through the impact layer lithology that fills these structures. At the Starkville North site, the contact between the boundary claystone and the impact layer has the appearance of an erosion surface, and the relief along the contact is as much as several millimeters (Fig. 13B). Thus, the evidence seems to suggest that the top of the boundary claystone was a paleosurface upon which plants grew (Figs. 1 and 4) and the impact layer was deposited. The amount of time represented by this paleosurface or diastem must be geologically brief.

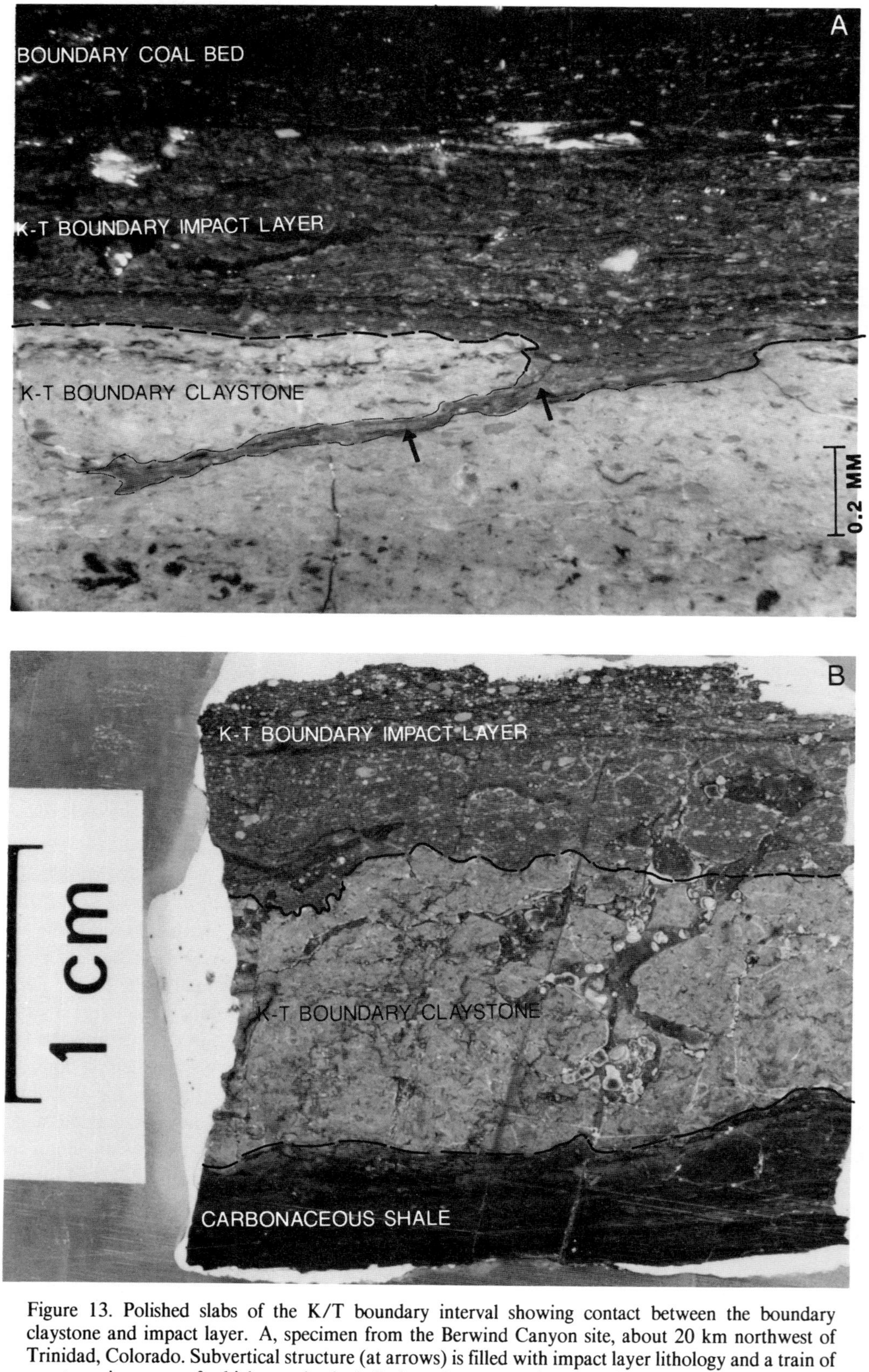

Figure 13. Polished slabs of the K/T boundary interval showing contact between the boundary claystone and impact layer. A, specimen from the Berwind Canyon site, about 20 km northwest of Trinidad, Colorado. Subvertical structure (at arrows) is filled with impact layer lithology and a train of quartz grains some of which are shocked. B, irregular contact between the boundary claystone and impact layer at the Starkville North site, a few kilometers south of Trinidad, Colorado.

TABLE 2. X-RAY FLUORESCENCE ANALYSES OF THE K/T BOUNDARY CLAYSTONE BED FROM COLORADO, NEW MEXICO, MONTANA, AND TONSTEINS FROM KENTUCKY*

	SVS	SVS	SVN	CCN	MADE	RAT	MR	BER	BB	BB	P6	OH
SiO_2	47.1	46.7	45.3	46.7	46.2	44.7	45.9	45.0	48.7	49.2	40.7	44.5
Al_2O_3	34.4	35.1	35.7	35.5	35.7	36.0	35.6	36.3	31.7	29.7	35.5	38.2
Fe_2O_3	0.62	0.67	1.23	0.68	0.60	1.61	1.37	1.17	0.88	1.20	7.18	0.56
MgO	0.37	0.37	0.53	0.45	0.37	0.44	0.29	0.40	0.49	0.70	0.79	0.27
CaO	0.18	0.14	0.30	0.17	0.20	0.22	0.12	0.27	0.06	0.04	0.09	0.1
Na_2O	0.17	<0.15	<0.15	<0.15	<0.15	<0.15	0.16	<0.15	0.96	1.23	<0.15	<0.15
K_2O	0.46	0.44	0.35	0.28	0.36	0.25	0.30	0.27	0.12	0.17	0.38	0.42
TiO_2	1.76	1.86	1.80	1.68	1.92	1.64	1.59	1.91	1.37	1.65	1.68	2.27
P_2O_5	0 06	0.06	0.11	0.10	<0.05	0.11	0.08	0.10	<0.05	<0.05	0.09	<0.05
MnO	<0.02	<0.02	<0.02	<0.02	<0.02	<0.02	<0.02	<0.02	<0.02	<0.02	<0.02	<0.02
LOI	15.0	14.9	14.9	14.5	15.1	15.4	15.1	15.0	15.0	15.7	13.0	13.8

*Analysts: A. J. Bartel, K. C. Stewart, and J. E. Taggart, U.S. Geological Survey; elements reported as oxides in weight percent; Fe calculated as Fe_2O_3; LCI, lost on ignition at 900°C. SVS, Starkville South (Colo.); SVN, Starkville North (Colo.); CCN, Clear Creek North (Colo.); MADE, Madrid East (Colo.); RAT, Raton (N. Mex.); MR, Madrid Railroad (Colo.); BER, Berwind Canyon (Colo.); BB, Brownie Butte (Mont.); P6, tonstein in Princess 6 coal bed (Ky.); OH, Olive Hill flint clay (Clack Mt., Ky.); Figure 2 shows location of sites in Raton basin.

Chemical composition of the K/T boundary claystone

Chemical analyses of boundary claystone samples were made by four different analytical methods to obtain a varied and extensive picture of its major- and trace-element composition. The methods employed were x-ray fluorescence (XRF), induction-coupled plasma atomic emission spectroscopy (ICP), instrumental neutron activation analysis (INAA), and radiochemical neutron activation analysis (RNAA).

XRF analyses show that the major-element content of the claystone (~95 percent SiO_2 and Al_2O_3 calculated on a water-free basis) closely resembles that of the clay mineral kaolinite. Small differences in the amounts of iron, magnesium, sodium, potassium, calcium, titanium, manganese, and phosphorus occur in claystone samples from place to place in the Raton basin (Table 2). The volatile content, mostly water, is about 15 percent. The amount of titanium, about 2.0 percent, calculated water-free as TiO_2, is slightly more than that of kaolinitic tonstein layers in coals of the Raton basin.

Analyses of the boundary claystone from Brownie Butte, Montana, are listed in Table 2. The claystone at the Montana site is similar to the boundary claystone bed in the Raton basin except that the Montana material contains more sodium and less potassium. This data is compatible with x-ray diffraction data, which indicate that the claystone is more smectitic than the Raton basin boundary claystone. Also, the Montana claystone contains slightly less titanium than does the Raton basin claystone.

Gilmore and others (1984) considered the amount of titanium in the boundary claystone to be an important tool for its recognition. This idea is confirmed by the data of Table 2. The titanium-to-aluminum ratio of 10 samples is 0.047 ± 0.005 (1σ). They concluded that the progenitor of the boundary claystone was about 97 percent impact target rock of basaltic composition. Inherent in their reasoning was the assumption that the titanium-to-aluminum ratio remained constant during diagenesis of the progenitor basalt (glass?). This assumption may not be warranted because titanium was mobile during diagenesis of boundary-interval sediments inasmuch as authigenic titanium oxide spherules occur in the boundary claystone at the Sussex, Wyoming, site. Moreover, authigenic goyazite spherules in the boundary claystone in Colorado and Wyoming contain significant amounts of titanium.

Spears and Kanaris-Sotiriou (1979) also assumed that the titanium-to-aluminum ratio remains constant during diagenesis of volcanic glass to clay. The ratio of titanium to aluminum in the boundary claystone (0.047) is greater than that of most European tonsteins (0.004 to 0.04) considered by Spears and Kanaris-Sotiriou (1979, Fig. 4) to be composed of altered silicic ash. Moreover, the titanium-to-aluminum ratio in the boundary claystone (0.047) is significantly less than that in other European tonsteins considered by Spears and Kanaris-Sotiriou (1979, Fig. 4) to be composed of altered basaltic ash (>0.16). If this were true, the boundary claystone is not altered silicic or basaltic volcanic ash because its titanium-to-aluminum ratio is markedly different than either of these compositional types. However, it is plausible that the boundary claystone could have formed from intermediate (andesitic) composition volcanic glass, if a volcanic origin for the boundary claystone were invoked.

Although Gilmore and others (1984) stated that the composition of the boundary claystone is different than tonsteins they studied in the Raton basin, that is not to say that it differs from other tonsteins or underclays elsewhere in the world. For example, Huddle and Patterson (1961) pointed out that kaolinitic underclays below coal beds commonly contain from 0.5 to 1.5 percent titanium expressed as TiO_2, and the titanium-to-aluminum ratio might fall in the range of the boundary claystone.

An analysis of a Pennsylvanian-age tonstein in the Princess 6 coal bed of Lawrence County, Kentucky (Table 2, P6), shows that it has a titanium-to-aluminum ratio of 0.047, identical to the boundary claystone. An analysis of the Olive Hill flint clay of Pennsylvanian age at Clack Mountain, Kentucky (Table 2, OH) shows that it has a titanium-to-aluminum ratio of 0.059, similar to that of the K/T boundary claystone. D. M. Triplehorn of the University of Alaska (oral communication, 1989) believes that both of these claystones (tonsteins) are composed of altered silicic volcanic ashes that contain beta quartz crystals.

The question of the relative mobility of major and trace elements during the diagenetic conversion of volcanic glass to clay was studied by Zielinski (1982). He compared the upper glassy part of a Miocene ash bed in the Troublesome Formation of Colorado mapped by Izett and Barclay (1973) with the lower part, which had been converted to clay (smectite). During devitrification of the glass to clay, magnesium, calcium, iron, titanium, manganese, phosphorous, and aluminum were enriched and potassium, sodium, and silicon were depleted. The titanium-to-aluminum ratio remained nearly, but not exactly, constant. Among the trace elements, cesium, rubidium, uranium, and zirconium were depleted in the clay-altered ash (smectite) relative to the glass during diagenesis. Strontium, scandium, chromium, and cobalt were enriched in the smectite relative to the glass. In general, the rare-earth elements (REE) were leached from the glass during its conversion to clay.

Zielinski (1985) also compared chemical analyses of 979 silicic upper Cenozoic volcanic ash beds, many of which had been collected and studied by the present writer, with analyses of a single Eocene kaolinitic tonstein at six sites in the Powder River basin of Wyoming. He showed that, with the exception of aluminum and possibly titanium, the tonsteins were depleted during diagenesis of major rock-forming elements relative to the silicic ashes. Among the trace elements, rubidium and cesium were depleted, transition-series metals were moderately depleted, and gallium, zirconium, and hafnium were relatively immobile. Some elements were depleted, including thorium, tantalum, niobium, yttrium, and the REE, which are generally considered immobile during low-temperature diagenesis.

Amounts of the REE in the boundary claystone from the Raton basin and Montana, as determined by the INAA method, are listed in Table 3. Their amounts were normalized to an average of chondrites (Boynton, 1984, Table 3.3) and plotted on Figure 14. The plots show that the REE content of the claystone is low; lanthanum is only enriched ten times and lutetium about twice relative to the average of the chosen chondrites. Both patterns show a small negative europium anomaly. The REE content of the boundary claystone from Western Interior North American sites is an order of magnitude less than the average North American shale reported by Haskin and others (1968). Although the amounts of the REE in the boundary claystone are less than that in average North American shale, the shape of the REE pattern of the boundary claystones is similar to that of average North American shale and to the shapes of kaolinite (Fig. 14) and smectite REE patterns published by Cullers and others (1975). Moreover, the shape of the REE pattern of the boundary claystone from the Raton basin is similar to that of clay in the K/T boundary interval at Stevns Klint, Denmark, and Gubbio, Italy, presented by Kastner and others (1984), except that the patterns of the boundary claystone from the Raton basin and Montana have significantly lower chondrite-normalized values (Fig. 14).

Zielinski (1982) showed that in some cases the REE were leached from silicic volcanic ash during its alteration to smectite, but the general shape of the chondrite-normalized-REE pattern remained similar to the parent glassy ash. In contrast, the chondrite-normalized-REE patterns of some altered silicic volcanic ash beds are markedly different than those of normal silicic volcanic ash beds (G. A. Izett, unpublished data). Kendrick (1985) showed that ashed samples of coal have very low REE contents, about 10 to 100 times that of chondrites. He conjectured that the REE are mobile during the alteration of volcanic glass to kaolinitic tonsteins.

Hildebrand and Boynton (1987) assumed that the K/T boundary claystone in the Raton basin is composed of altered impact material. They further assumed that the REE in the claystone were immobile during diagenesis, and they concluded, therefore, that the REE data for the claystone could provide a

TABLE 3. INSTRUMENTAL NEUTRON ACTIVATION ANALYSES OF THE K/T BOUNDARY CLAYSTONE BED FROM COLORADO AND MONTANA*

Element	SVS	BB
Cs	0.76	0.27
Rb	12.25	4.53
Sr	82.9	40.3
Th	3.7	1.39
U	2.01	2.60
La	3.35	3.30
Ce	7.74	5.04
Nd	1.76	1.75
Sm	0.57	0.48
Eu	0.15	0.10
Tb	0.11	0.08
Yb	0.32	0.34
Lu	0.05	0.06
Ta	0.73	0.68
Hf	2.83	1.77
Sb	3.54	1.27
Sc	19.8	14.4
Mn	8.4	26.32
Fe	0.4	0.64
Cr	59.4	43.6
Co	1.72	2.93

*Analysts: J. R. Budahn and R. J. Knight, U.S. Geological Survey; Fe in weight percent; other elements in parts per million; SVS, Starkville South (Colo.); BB, Brownie Butte (Mont.); location of sites shown on Figures 2 and 45.

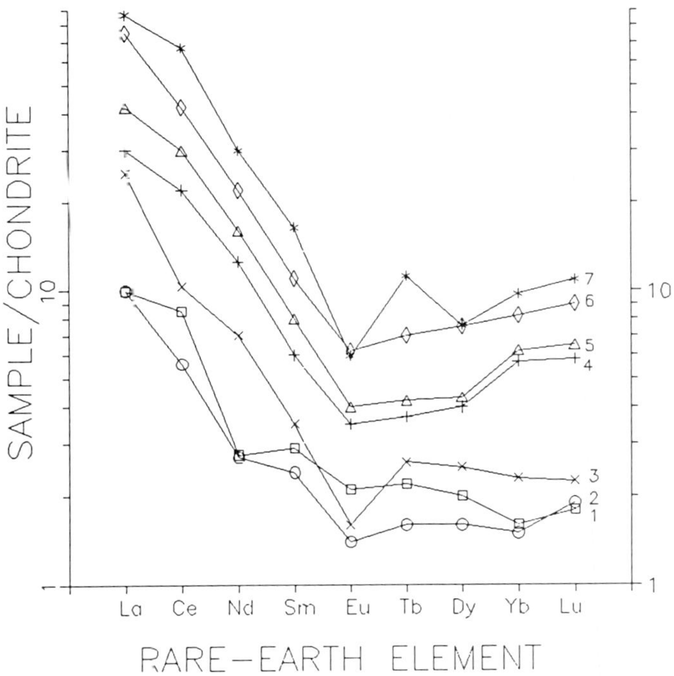

Figure 14. Rare-earth-element (REE) diagram of K/T boundary interval rocks from North America and Europe. A typical kaolinite pattern from Georgia is also shown. Element concentrations normalized to amounts in chondritic meteorites plotted against REE atomic number. 1, boundary claystone from the Starkville South K/T boundary site, Colorado; 2, boundary claystone at Brownie Butte, Montana; 3, boundary claystone residue from Stevns Klint, Denmark (Kastner and others, 1984); 4, insoluble residue of Tertiary clay directly overlying K/T boundary at Stevns Klint, Denmark (Kastner and others, 1984); 5, insoluble residue of Cretaceous clay directly underlying K/T boundary at Stevns Klint, Denmark (Kastner and others, 1984); 6, insoluble residue from K/T boundary claystone at Gubbio, Italy (Kastner and others, 1984); 7, Macon, Georgia, kaolinite of Cullers and others (1975).

basis for ascertaining the composition of the progenitor of the claystone. The fact that the boundary claystone has an unusually low REE content led them to speculate that the progenitor of the boundary claystone was composed of a mixture of depleted oceanic mantle, oceanic crust, and continental sediments. Their assumption that REE are immobile during diagenesis is not in agreement with the fact that some REE are relatively abundant (cerium about 300 ppm) in authigenic goyazite spherules in the K/T boundary claystone. This fact indicates that the REE were indeed mobile during diagenesis of boundary interval rocks. Therefore, the REE signature may not reflect the chemical nature of the progenitor material of the boundary claystone.

Chemical analyses by the ICP method for 17 trace elements in the boundary claystone are listed in Table 4. For comparison, analyses by the same analytical method of carbonaceous shale, claystone, coal, and jarosite from the K/T boundary interval are listed in Table 5. The analyses (Tables 4 and 5) show that the composition of the boundary claystone, for the elements determined, is similar to the "average shale" of the Earth's crust (Turekian and Wedepohl, 1961). The amounts of the REE in the boundary claystone and carbonaceous claystone from 1 cm above the K/T impact layer are similar. Both have lanthanum contents only about ten times that in chondrites.

Gilmore and others (1984) observed that the chemical composition of the boundary claystone is slightly different than tonstein layers in coal beds of the Raton basin, and their observation is confirmed by the analyses in Table 4 of this paper. They (1984, p. 4) emphasized that the "enhanced" amount of titanium, chromium, and scandium in the boundary claystone suggests that the claystone is formed of altered basaltic material, perhaps basaltic volcanic ash. This conclusion may not be correct because the amounts of these elements are no different than that of average shale. Furthermore, the claystone lacks calcic plagioclase phenocrysts and microlites that commonly occur in basaltic ash. Although the chemical composition of the boundary claystone may be slightly different than that of Cretaceous tonsteins of the Raton basin, that is not to say that its composition differs (except platinum-group metals) from the average North American shale.

Of interest is the low nickel (2 to 5 ppm) and cobalt (1 to 2 ppm) content of the K/T boundary claystone (Table 4) in the Raton basin. The low content of nickel and cobalt is surprising because some (Pollastro and Pillmore, 1987) believe that the claystone consists of altered impact ejecta, target and asteroid, the asteroidal component of which should contain significant amounts of nickel and perhaps cobalt. In contrast to the low content of these elements (<5 ppm) in the boundary claystone in continental K/T boundary rocks of the Raton basin, the boundary impact layer in marine rocks at Caravaca, Spain, and Stevns Klint, Denmark, contains 2,580 (ppm) and 1,370 (ppm) nickel and 626 (ppm) and 146 (ppm) cobalt, respectively (Kyte and others, 1985). The occurrence of large amounts of nickel and cobalt in marine K/T boundary rocks has been used as evidence to support the K/T impact origin of the claystone. The low content of nickel, in particular, in the continental K/T boundary rocks of the Raton basin suggests that the boundary claystone and impact layer essentially do not contain any asteroidal material. Another explanation for the low content of these elements is that they were severely and equally depleted from K/T boundary interval rocks during diagenesis.

The iridium, silver, and gold contents of the boundary claystone from four sites in the Raton basin were determined by H. T. Millard, Jr., of the U.S. Geological Survey, by the RNAA method, and the data are listed in Table 6 and illustrated on the stratigraphic sections of Figure 15. In general, the iridium data are similar to those reported by Orth and others (1981), Pillmore and others (1984), and Gilmore and others (1984). However, the iridium data sets are not strictly comparable because these workers did not recognize or sample the impact layer as a separate microstratigraphic unit; accordingly, some of their samples may have been mixtures of the impact layer and the boundary claystone. Millard's analyses indicate that the boundary claystone contains generally less than 1.0 ppb; in contrast, the impact layer contains as much as 66 times more iridium than the boundary claystone (Fig. 15).

TABLE 4. INDUCTION-COUPLED ATOMIC EMISSION SPECTROSCOPIC ANALYSES OF THE K/T BOUNDARY CLAYSTONE BED FROM COLORADO, NEW MEXICO, AND MONTANA*

Element	SVN	SVS	CCN	RAT	MAD	MADE	BER	BB
Ti	1.0	1.1	1.0	1.1	0.9	1.2	1.1	0.8
Mn	5	4	<17	15	5	<4	5	25
As	50	<10	20	120	40	40	20	50
Ba	210	150	180	370	650	320	260	120
Ce	78	17	62	80	81	7	76	10
Co	2	1	1	1	1	2	<1	2
Cr	65	70	51	49	39	52	68	53
Cu	12	3	6	19	7	11	69	9
Ga	30	31	30	32	31	35	32	<2
La	37	6	27	43	39	2	40	4
Li	220	280	240	240	250	200	180	200
Mo	7	8	9	12	6	13	6	2
Nd	38	5	26	28	40	<4	32	<4
Ni	4	3	4	4	<2	5	2	15
Sc	19	20	17	19	15	20	23	13
V	120	140	110	130	86	110	130	82
Zn	17	15	28	17	12	17	12	26

*Analyst: M. J. Malcolm, U.S. Geological Survey; Ti in weight percent; other elements in parts per million. SVN, Starkville North (Colo.); SVS, Starkville South (Colo.), CCN, Clear Creek North (Colo.); RAT, Raton (N. Mex.); MAD, Madrid (Colo.); MADE, Madrid East (Colo.); BER, Berwind Canyon (Colo.); BB, Brownie Butte (Mont.); location of sites shown on Figures 2 and 45.

Authigenic spherules

In the Raton basin, the K/T boundary claystone contains three different compositional types of authigenic spherules, including (1) kaolinite, (2) goyazite, and (3) two iron-rich minerals. Kaolinite and goyazite spherules also are found in K/T boundary rocks elsewhere in western North America. In addition, rare authigenic titanium oxide (probably rutile) spherules were found in the boundary claystone at the Sussex, Wyoming, K/T boundary site. Kaolinite and goyazite spherules occur only in the boundary claystone and not in the overlying impact layer.

The spherules were studied on the surfaces of polished slabs, in thin sections, and in oil-immersion mounts using optical methods. In addition the spherules were studied using an SEM. Because the spherules occur in the boundary claystone, thought by some (Pollastro and others, 1983) to be composed of altered glassy impact material, the spherules also were interpreted to be altered impact material analogous to microtektites (Smit and Romein, 1985, p. 156). However, I will present evidence that demonstrates that the spherules are authigenic.

Kaolinite spherules are always solid; their surfaces are finely pitted, reflecting the microspherulitic texture of the surrounding kaolinite host. The spherules range in diameter from 0.1 to 0.5 mm. They consist of a kaolinite core encased in a 5- to 10-μm-thick shell or rim of kaolinite having a different habit than the core (Fig. 16). The texture of the cores of the spherules, as seen using an optical petrographic microscope, is a uniform mosaic of kaolinite granules a few micrometers in diameter. SEM study of the spherules showed that the kaolinite cores are composed of an amazingly uniform polygonal framework of kaolinite filled with kaolinite spherules 2 to 4 μm in diameter. This texture is similar to, but more regular than, that of the kaolinitic matrix of the enclosing boundary claystone (Fig. 16). As seen in thin sections, the outermost region of the kaolinite shells contains a concentration of pale brown submicrometer-size inclusions, possibly a titanium-rich mineral.

SEM-EDS study of the shells of the spherules showed that they are composed of columnar kaolinite that contains considerably more titanium than the kaolinite of the boundary claystone host or the cores of the spherules. The contact between the columnar shell and microspherulitic core of the kaolinite spherules is sharp, as shown in Figure 16. The detailed structure of such kaolinite spherules is similar to that of authigenic glauconite pellets described by Odom (1976, Fig. 3). Both have a microgranular core surrounded by a thin rim or shell.

Kaolinite spherules have not been found at K/T boundary sites in western North America where the claystone is mainly composed of smectite. For example, no kaolinite spherules were seen on polished surfaces of the boundary claystone from the Frenchman River site in southwestern Saskatchewan and the Red Deer Valley site in Alberta.

Spherules composed of a mineral here referred to as goyazite

TABLE 5. INDUCTION-COUPLED ATOMIC EMISSION SPECTROSCOPIC ANALYSES OF ROCKS FROM THE K/T BOUNDARY INTERVAL FROM COLORADO AND NEW MEXICO*

Element	SVS Shale-1	SVS Shale-2	BER Clay	BER Coal	CCN Jarosite	Crust Average Shale
Ti	0.8	1.2	0.50	0.04	0.3	0.46
Mn	17	14	40	8	20	850
As	80	20	10	10	1500	13
Ba	380	360	380	47	150	580
Ce	29	9	82	19	62	60
Co	9	11	4	9	4	19
Cr	240	230	68	23	120	90
Cu	23	28	62	24	37	45
Ga	24	21	29	4	7	19
La	10	4	46	9	33	30
Li	25	21	35	2	13	66
Mo	6	4	<2	<4	<4	11
Nd	12	5	35	12	24	28
Ni	13	15	10	68	<2	68
Sc	28	37	21	16	10	13
V	220	180	170	90	360	130
Zn	58	61	42	78	45	95

*Analyst: M. J. Malcolm, U.S. Geological Survey; Ti in weight percent; other elements in parts per million. Shale-1, millimeter-thick dark flaky shale directly overlying K/T boundary impact layer and underlying coal bed; Shale-2, carbonaceous claystone about 1 cm above K/T boundary impact layer; Clay, claystone about 4 cm below K/T boundary claystone; Coal, K/T boundary coal bed; Jarosite, jarosite laminae in K/T boundary impact layer; Crust, analysis of average shale of Earth's crust reported by Turekian and Wedepohl (1961). SVS, Starkville South (Colo.); BER, Berwind Canyon (Colo.); CCN, Clear Creek North (Colo.); location of sites shown on Figure 2.

are widely dispersed in the K/T boundary claystone in the Raton basin. Goyazite is the strontium-rich member of an isostructural group of several hydrous aluminum phosphate minerals. The group of minerals is named for the lead-rich member of the series, plumbogummite. Gorceixite and crandallite are the barium- and calcium-rich members, respectively. Although the mineral has been referred to as goyazite (Bohor and others, 1987; Pillmore and Flores, 1987), its cation content is mainly barium and subequal amounts of strontium and calcium.

F. E. Lichte and I made laser-ablation inductively coupled mass spectrometric (LA-ICP-MS) analyses of the aluminum phosphate spherules. This method permits the detection, in the parts per billion range, of about 75 elements during one series of laser pulses. Analysis of the aluminum phosphate mineral showed that it contains 7.2 percent barium, 5.2 percent calcium, and 4.4 percent strontium. Thus, the LA-ICP-MS analytical data indicate the mineral is more like gorceixite than goyazite, but it is herein referred to as goyazite for convenience only.

The LA-ICP-MS analyses also showed that the goyazite contains iron (4.5 percent, as Fe_2O_3), titanium (0.74 percent, as TiO_2), lanthanum (115 ppm), cerium (280 ppm), and nickel (4.2 ppm). All platinum-group metals were looked for in the goyazite, but were not detected at the 8 ppb level.

Aluminum phosphate minerals similar to goyazite occur in kaolinitic tonstein layers that in some instances are associated with coal beds (Triplehorn and Bohor, 1983). Price and Duff (1969) pointed out that goyazite occurs in certain tropical soils, particularly in laterites. Recently, Triplehorn and Finkleman (1989) found microspherules (also teardrop-shaped objects) composed of a calcium- and strontium-bearing aluminum phosphate mineral (probably goyazite) in a tonstein of volcanic origin in the Cannel City coal bed of Pennsylvanian age in eastern Kentucky. D. M. Triplehorn of the University of Alaska (written communication, 1989) also found spherules composed of alumi-

TABLE 6. RADIOCHEMICAL NEUTRON ACTIVATION ANALYSES FOR IRIDIUM, GOLD, AND SILVER IN THE K/T BOUNDARY CLAYSTONE AND IMPACT LAYER FROM COLORADO AND NEW MEXICO*

Element	SVS K/T	CCN K/T	RAT K/T	MADE K/T	SVS IMP	CCN IMP	RAT IMP	MADE IMP
Ir	0.27	0.22	0.07	0.32	1.21	14.6	1.2	7.5
Au	1.1	2.0	4.4	5.1	1.7	7.3	2.5	9.4
Ag	31	74	100	56	50	280	63	70

*Analyst: H. T. Millard, Jr., U.S. Geological Survey; all elements in parts per billion; K/T, K/T boundary claystone bed; IMP, K/T boundary impact layer. SVS, Starkville South (Colo.); CCN, Clear Creek North (Colo.); RAT, Raton (N. Mex.); MADE, Madrid East (Colo.); location of sites shown on Figure 2.

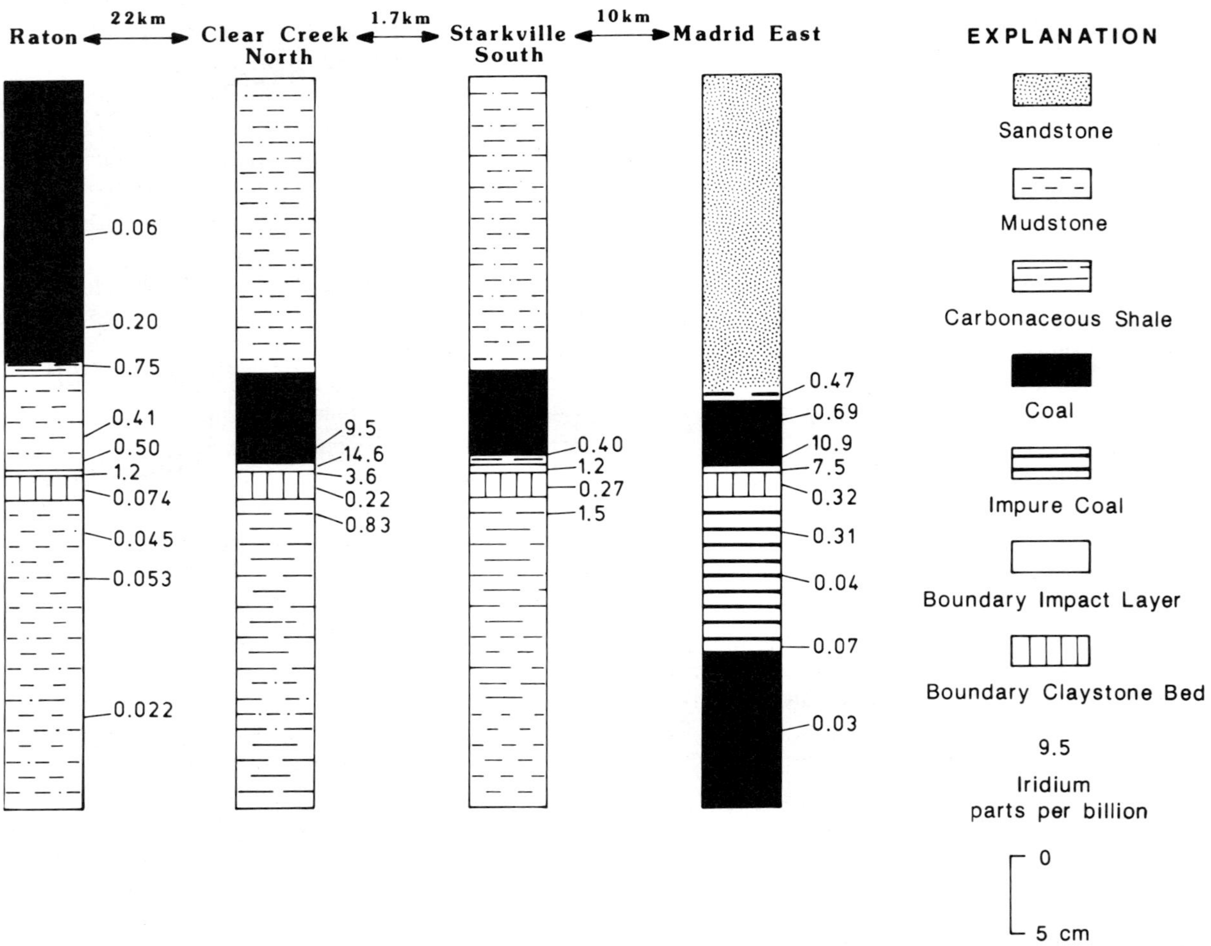

Figure 15. Stratigraphic diagram showing amounts of iridium (ppb) in K/T boundary rocks of the Raton basin. Locations of stratigraphic sections shown on Figure 2. Iridium analyses made by H. T. Millard, Jr., U.S. Geological Survey following methods he described in 1987.

num phosphate (probably goyazite) disseminated in the Wyodak coal bed of Paleocene age at the Buckskin mine in Wyoming.

Goyazite spherules in the boundary claystone in the Raton basin generally are light brown, opaque, and waxy appearing; others are colorless to reddish brown and translucent. SEM study of the spherules showed that many are amazingly smooth, but some have a finely pitted surface reflecting the microspherulitic texture of the surrounding kaolinite host. The spherules range from about 50 to 150 μm in diameter, but most are about 80 μm in diameter. Plainly, goyazite spherules are much smaller than coexisting kaolinite spherules, which typically are 100 to 500 μm in diameter. As will be pointed out in a following paragraph, goyazite spherules are much larger at Wyoming K/T boundary sites.

Izett (1987a, Fig. 9) pointed out that many goyazite spherules are nearly perfect spheres; however, some are oblate spheroids and others are compound and form intergrown groups and mammillary masses. Most are hollow and have thick rims; a few are solid and consist of radially oriented sheaf-like crystal masses. The internal surfaces of the shells of hollow spherules are commonly mammillary. Rarely, the exterior surfaces of spherules show impressions of a coauthigenic mineral, possibly pyrite or goyazite (Izett, 1987a, Fig. 9). The outermost rims of goyazite spherules are consistently enriched in barium, whereas the outermost rims of kaolinite spherules are enriched in titanium.

Most of the goyazite spherules in the boundary claystone in the Raton basin are small, thick walled, and hollow, but some are solid. Rare spherules are larger (0.2 mm) and consist of goyazite and kaolinite. Their cores are microcrystalline platy kaolinite, whereas the cores of kaolinite spherules are microspherulitic. The kaolinite cores are surrounded by a 10-μm-thick shell of goyazite (Fig. 18A). The interior surface of the goyazite shell is mammillary, and growth impressions of the goyazite are on the outer surface of the kaolinite core. This relation suggests that the goyazite shell formed first in a spherical cavity, and the cavity was later filled with platy kaolinite.

A third compositional type of spherule, also found in the boundary claystone in the Raton basin, consists of two different types of iron-rich minerals. One type forms porous, spherical masses 10 to 50 μm in diameter composed of jarosite. Such

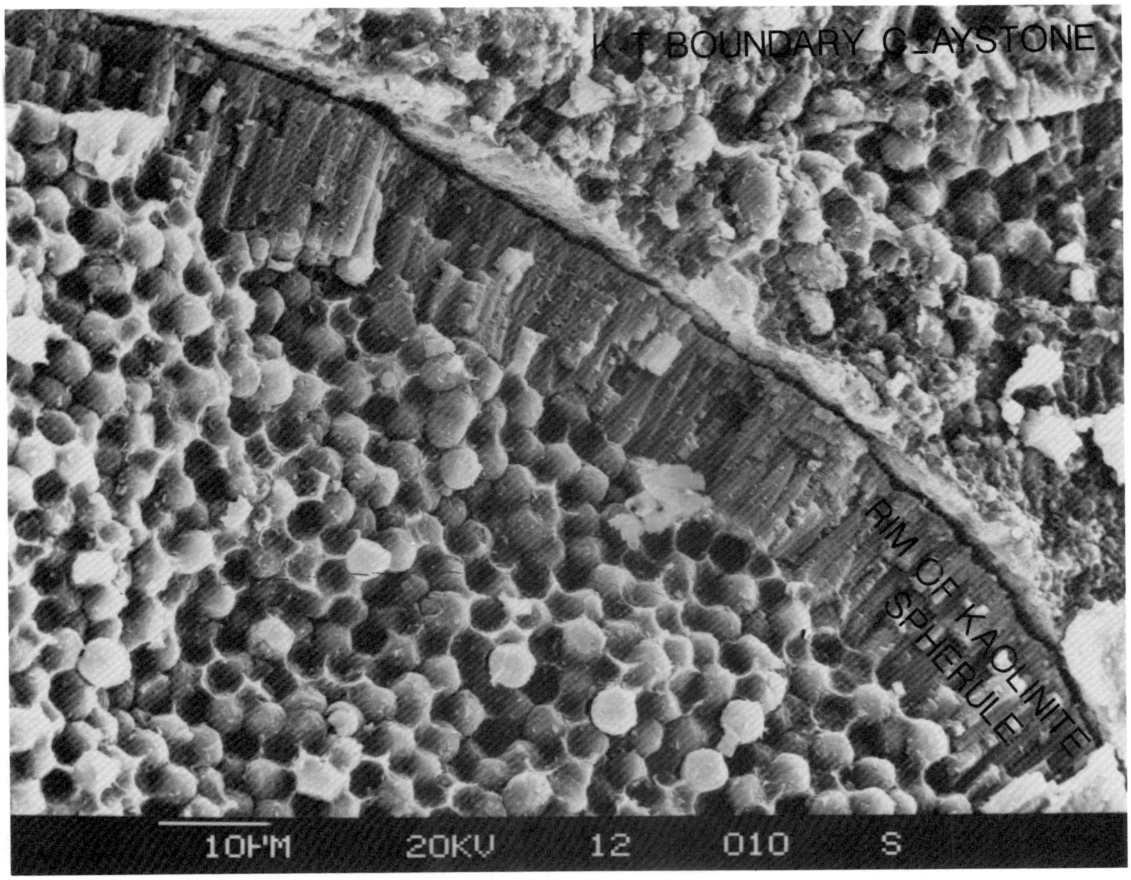

Figure 16. Kaolinite spherule in the K/T boundary claystone from the Clear Creek North site, south of Trinidad, Colorado. The rim of the spherule is columnar kaolinite whereas the core is composed of microspherules of kaolinite set in a honeycomb-like kaolinite framework. Texture within the microspherule is more regular than in the surrounding boundary claystone.

spherules also occur in the overlying impact layer. These spherules generally occur singly or in clusters in or near concentrations of carbonaceous material. A second type of iron-rich spherule is composed of a translucent, reddish brown, hydrous iron-oxide mineral (goethite?). Cross sections of such spherules were seen on polished surfaces of slabs of K/T boundary interval rocks from the Madrid East site near Madrid, Colorado. The spherules, which are minute (10 to 20 μm), occur in the uppermost part of the boundary claystone just below the contact with the overlying impact layer. These iron-rich spherules are unquestionably authigenic and probably formed from framboidal pyrite or marcasite. There is no evidence that would suggest that they are of impact origin.

Another compositional type of spherule was found in K/T boundary rocks at the Sussex, Wyoming, site. They are reddish brown, about 100 μm in diameter, and consist of a shell of a titanium-rich mineral (probably rutile) that surrounds a core of microspherulitic kaolinite. The spherules, which are rare, were found in a 3-mm-thick layer of pink boundary clay lithology. This unit occurs below a layer of gray claystone about 1.2 cm thick that contains disseminated quartz grains (20 to 120 μm in diameter), disseminated shreds of carbonaceous material (vitrinite), and clasts of boundary clay lithology about 10 cm thick. The presence of these constituents implies that the claystone is composed of reworked material. Locally, a unit a few millimeters thick of pink boundary clay lithology overlies the gray unit composed of reworked material. The upper unit of pink boundary clay lithology is overlain by the K/T impact layer.

The distribution of kaolinite and goyazite spherules within the K/T boundary claystone was determined by microscopic study of polished surfaces of the claystone. It is important to stress the fact that kaolinite and goyazite spherules can occur in the boundary claystone in close proximity, within a fraction of a millimeter of each other. Kaolinite spherules are, for the most part, randomly and evenly disseminated; however, as many as ten spherules were found tightly clustered within the claystone. Individual spherules within some clusters were crushed during a compaction event that must have occurred early in the diagenetic history of the sediments. Hollow and solid goyazite spherules are also randomly and evenly scattered through the boundary claystone. However, clusters of as many as 25 goyazite spherules can occur in some places in the claystone. Individual goyazite spherules in the clusters were not compacted as the kaolinite spherules were. The cause or origin of the unusual distribution (clustering) for both the kaolinite and goyazite spherules is unknown.

An accurate assessment of the absolute abundance of kaolinite and goyazite spherules at K/T sites in the Raton basin is difficult to make because of their low abundance. Although their abundance is extremely low, microscopic study showed that kaolinite and goyazite spherules are not equally abundant at all K/T sites. Kaolinite spherules are most abundant at the Clear Creek North site and least abundant at the Berwind Canyon site. A point-counter analysis of the polished surfaces of the boundary claystone was made to obtain an approximate idea of the abundances of the kaolinite spherules. The analysis showed that the boundary claystone typically contained about 10 kaolinite spherules per square centimeter at the Clear Creek North site, but only about 2 per cm^2 at the Berwind Canyon site. Goyazite spherules are rare at all sites, but they are relatively more abundant at the Raton, Madrid, and Berwind Canyon sites as compared to their abundance at the Clear Creek North site. The claystone at the Raton site contained only about one goyazite spherule per 10 cm^2. Clearly, kaolinite spherules are much more abundant than goyazite spherules.

In contrast to the paucity of goyazite spherules in the K/T boundary claystone in the Raton basin, they are much more abundant at two Wyoming K/T boundary sites: Dogie Creek and Teapot Dome. However, no goyazite spherules were found in the K/T boundary claystone at a third newly discovered site near Sussex, Wyoming (Nichols and others, 1990), about 45 km northwest of the Teapot Dome site. At the Brownie Butte, Montana, K/T boundary site 500 km north of the Dogie Creek, Wyoming, area, Bohor and others (1984) did not report the presence of goyazite spherules, and I did not see any goyazite spherules in thin sections or on polished slabs of the boundary claystone. Moreover, spherules were not seen on polished surfaces of the boundary claystone from sites in Alberta and Saskatchewan, Canada. Because the origin of the goyazite spherules is controversial and because of their local abundance at two Wyoming K/T boundary sites, a considerable effort was made to study them at these sites.

The distribution of goyazite spherules within discrete lenticular microstratigraphic units at the Teapot Dome and Dogie Creek sites is distinctly different than the distribution of spherules in the K/T boundary claystone in the Raton basin where they are generally widely disseminated (Izett, 1987b). At the Teapot Dome and Dogie Creek K/T boundary sites, goyazite spherules are concentrated in a stratiform goyazite-mineralized kaolinite layer at the top of the K/T boundary claystone (Fig. 17). The goyazite mineralized layer, which is generally 1.5 cm thick, contains goyazite-rich stratiform layers and goyazite-poor disseminated material. Within the layer, goyazite replaced plant material along bedding planes, and the rectangular-shaped object in Figure 18B may have been a twig. Goyazite also can be concentrated in a thin band at the contact between the goyazite-mineralized layer and the adjacent claystone. Locally, the goyazite-mineralized layer pinches and swells in thickness or abruptly terminates against kaolinite masses and pellets (Fig. 17B).

At the Dogie Creek site, goyazite spherules are concentrated in lenticular masses as thick as 4.0 cm. The goyazite-mineralized layer overlies a unit of water-transported, reworked boundary claystone. That the boundary claystone consists of reworked material is demonstrated by its abundant content of quartz and other mineral grains that are not shock metamorphosed. At the Teapot Dome site, rare goyazite spherules occur in claystone 1.0 cm below the base of the boundary claystone. The claystone contains disseminated unshocked quartz and other mineral grains as well as shreds and flecks of carbonaceous plant material.

Within the goyazite-mineralized layer, goyazite spherules constitute as much as 30 percent of boundary claystone at the Dogie Creek, Wyoming, site and as much as 10 percent of the claystone at the Teapot Dome site. Goyazite spherules occur not only as nearly perfect spheres but also rarely as oblate spheroids. More rarely, goyazite forms objects shaped like teardrops, peanuts, cigars, summer squash, and eggs (Fig. 19). Some of these odd-shaped goyazite objects are similar to forms of revolution and to some microtektites. Thin-section study showed that goyazite occasionally forms subspherical cavity fillings that have scalloped borders (Fig. 20B). These odd-shaped objects are clearly unlike the shapes of microtektites.

The diameter of goyazite spherules at the Teapot Dome and Dogie Creek sites ranges from about 0.04 to 1.4 mm. The mean diameter of 100 complete goyazite spherules plucked from the claystone at the Dogie Creek, Wyoming, site was determined to be 0.90 ± 0.17 mm. Measurement of goyazite spherule diameters in oil-immersion mounts and thin sections demonstrated that the largest goyazite spherules are generally hollow and thin walled; the smallest ones are thick walled or solid. Of possible significance to the origin of the spherules is the fact that most spherules at Wyoming K/T boundary sites (Dogie Creek and Teapot Dome) are considerably larger (an order of magnitude) than those in the Raton basin. Moreover, the spherules at Dogie Creek are somewhat larger than those at Teapot Dome.

One of the chief characteristics of the spherules is that their goyazite walls generally have a mammillary inner surface reflecting the spherical growth habit of the goyazite. Spheres and partial spheres of colloform goyazite project into the central spherical cavity. The spherules thus have the appearance of hollow geodes.

Thin-section study of the goyazite-mineralized claystone that contains the goyazite spherules showed that the shells or rims of the spherules are colloform (Fig. 21); that is, they consist of laminae upon laminae of concentrically oriented fibrous goyazite (Fig. 21B). Apparently, spherical growth of goyazite began from a phosphate-rich colloidal fluid at point sources along the surface of spherical cavities. Some goyazite laminae are probably slightly different in chemical composition from their neighbors. This conclusion is based on the optical microscopic observation that some goyazite laminae have a slightly darker color as viewed in plane-

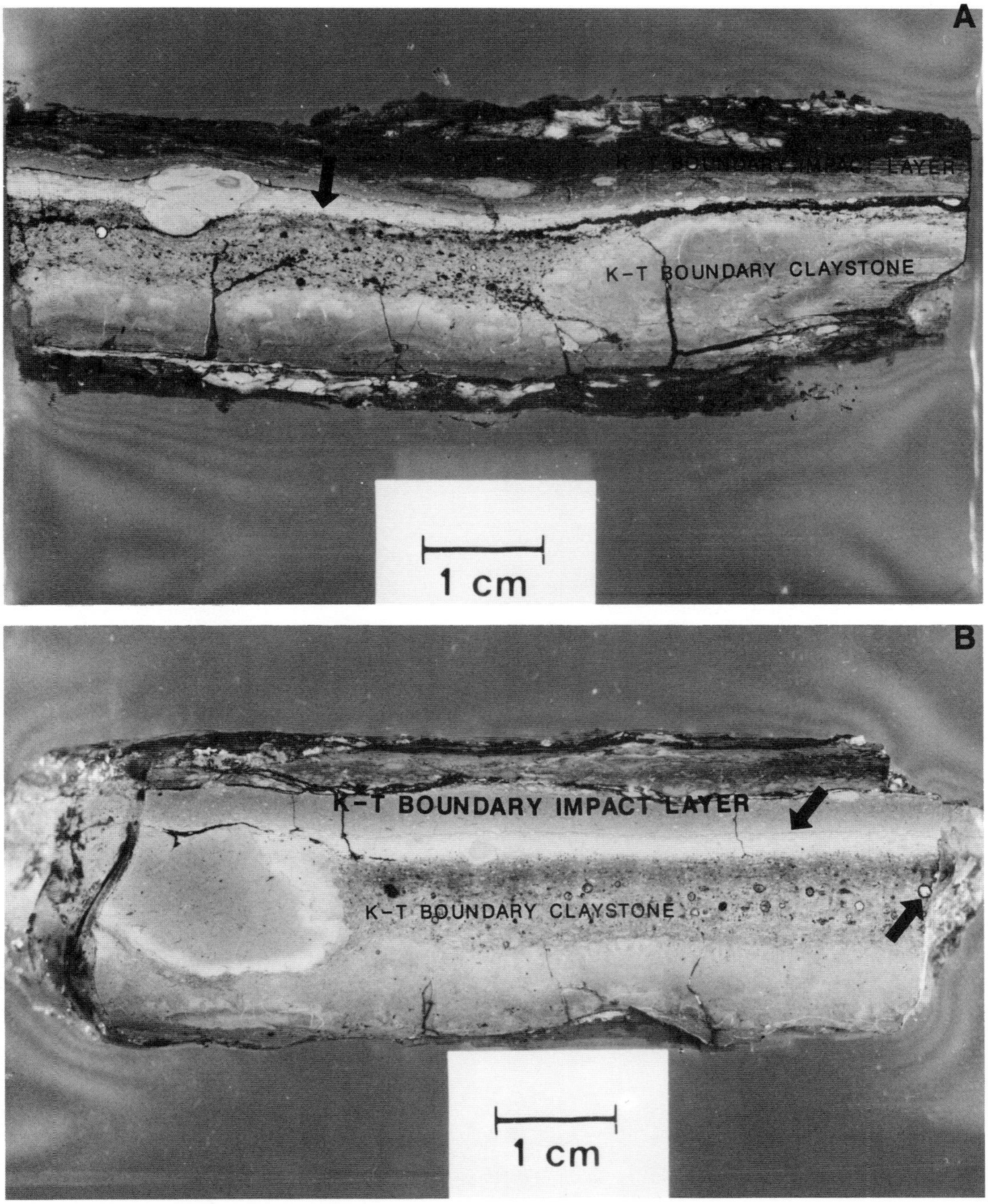

Figure 17. Polished slabs of the K/T boundary interval from the Teapot Dome, Wyoming, site. A, prominent dark layer that contains goyazite spherules occurs at the top of the K/T boundary claystone bed. First appearance of shocked mineral grains (at arrow) occurs on top of the goyazite layer. Prominent layer of goyazite-mineralized claystone abuts a region of kaolinitic claystone that is free of goyazite. B, prominent layer of goyazite-mineralized claystone abuts a pellet of kaolinitic claystone that is free of goyazite. Arrows point to K/T boundary impact layer and a large goyazite spherule.

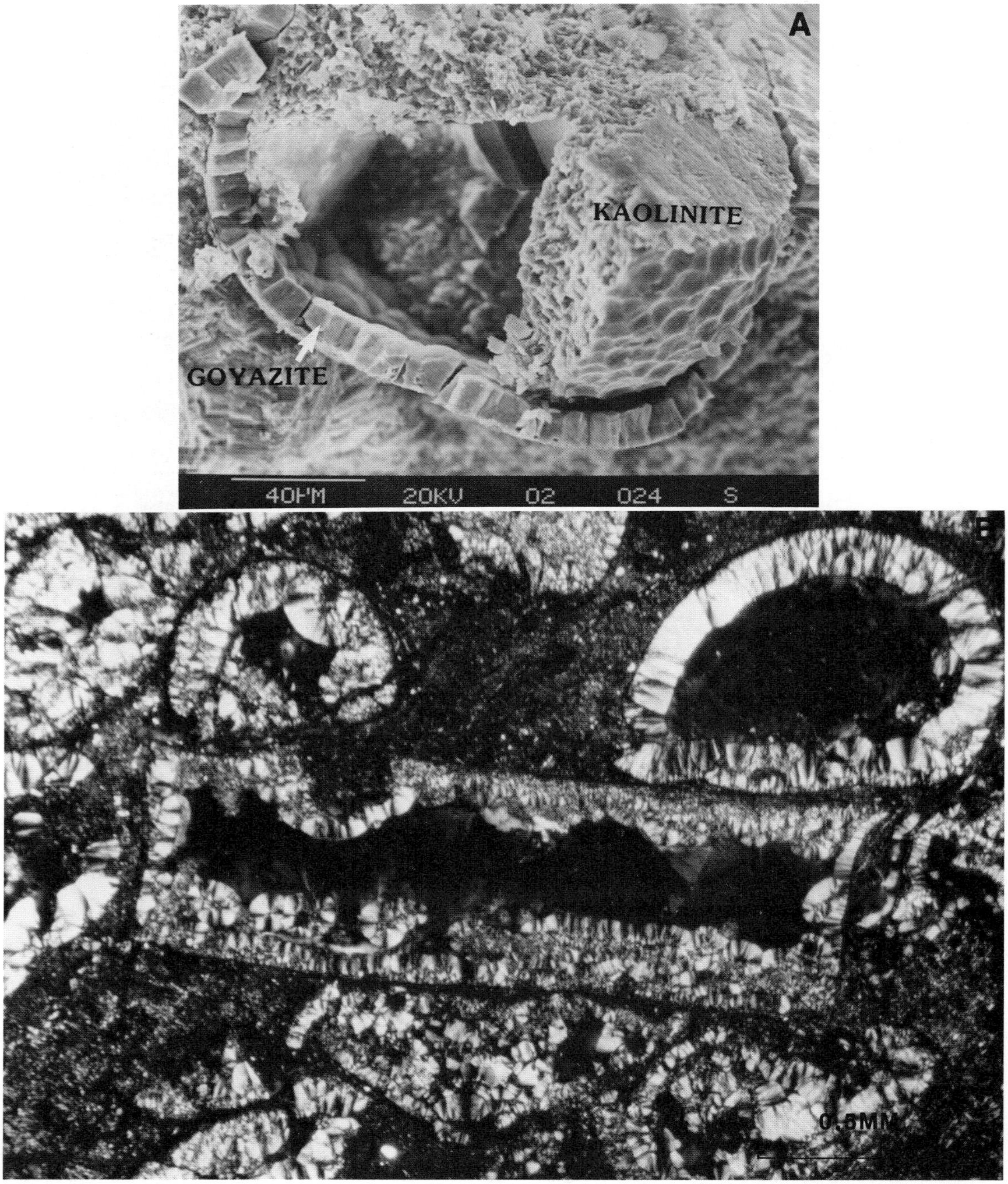

Figure 18. Goyazite and goyazite-kaolinite "spherules" from the K/T boundary claystone bed. A, SEM photomicrograph of "spherule" from the Madrid, Colorado, site. Shell of goyazite surrounds core of platy kaolinite. Outer surface of the kaolinite core has growth impressions of the mammillary inner surface of the goyazite shell. B, photomicrograph of thin section of the boundary claystone from Dogie Creek, Wyoming (Bohor and others, 1987). Goyazite occurs in crushed masses, deformed spherules, and a cavity filling that has a rectangular shape in cross section.

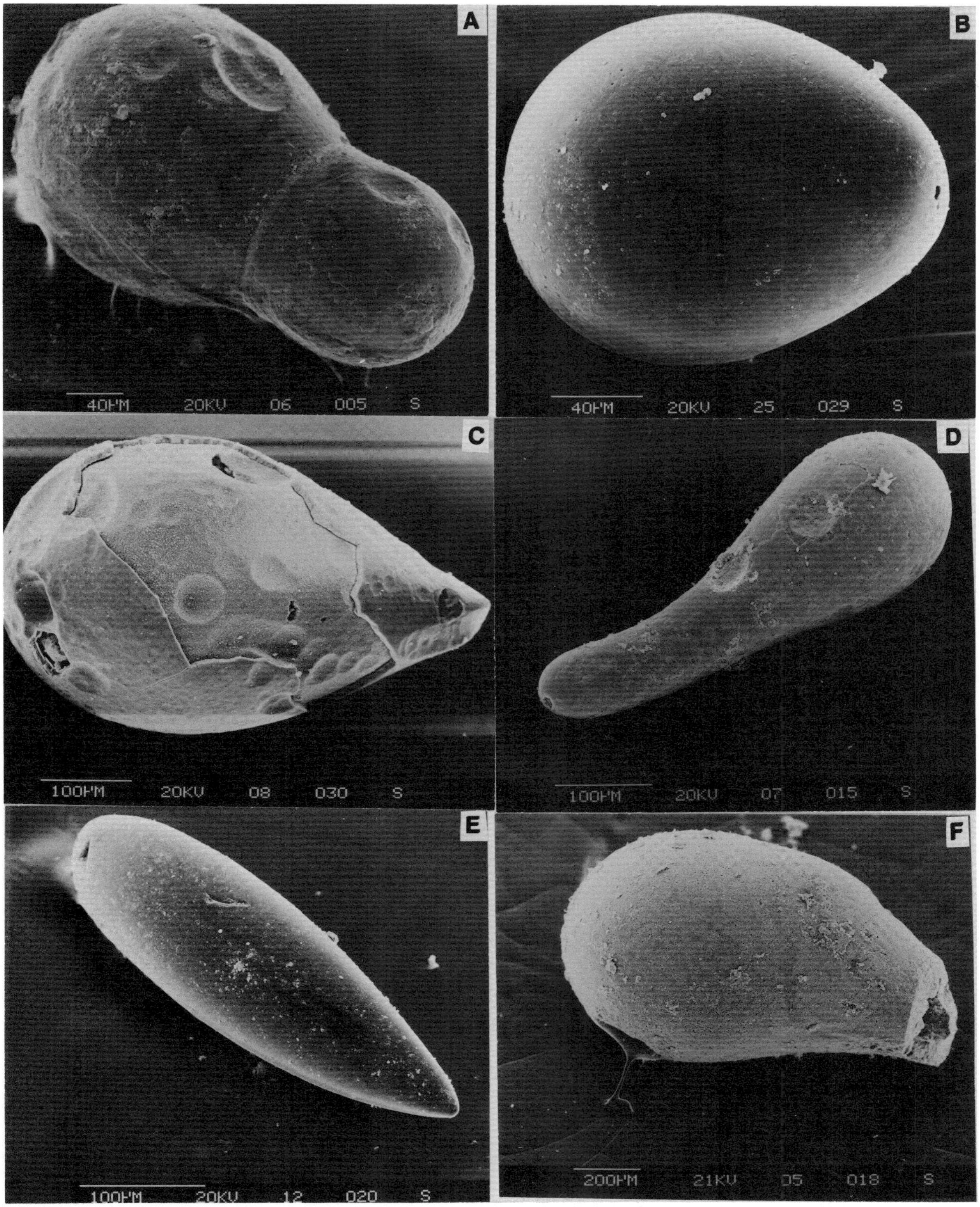

Figure 19. Rare types of goyazite objects loosely referred to as "spherules" in the K/T boundary claystone bed at the Dogie Creek, Wyoming, site. Among the goyazite spherules are: A, peanut-shaped goyazite object having dimpled surface; B, egg-shaped hollow goyazite object; C, teardrop-shaped object having dimpled surface and subsurface; D, summer squash-shaped goyazite object; E, cigar-shaped goyazite object; F, teardrop-shaped goyazite object having wrinkled surface.

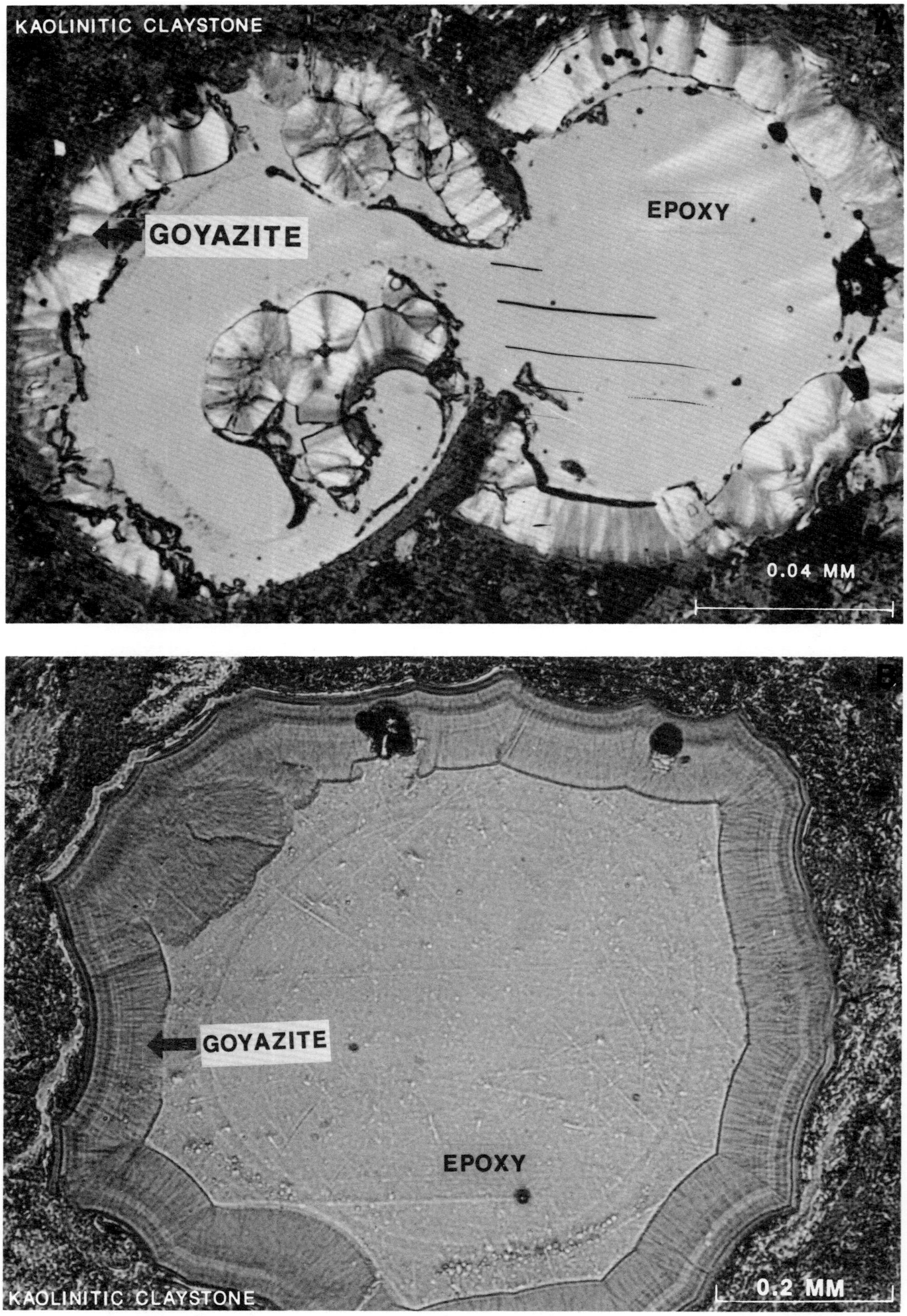

Figure 20. Hollow goyazite objects in the K/T boundary claystone from the Dogie Creek, Wyoming, K/T boundary locality. A, double spherule composed of colloform microlaminated goyazite. Either spherule on right formed first because its goyazite shell is cut off by the wall of the spherule to the left or the spherules were crushed. B, goyazite object showing a scalloped border. Wall of object composed of colloform goyazite.

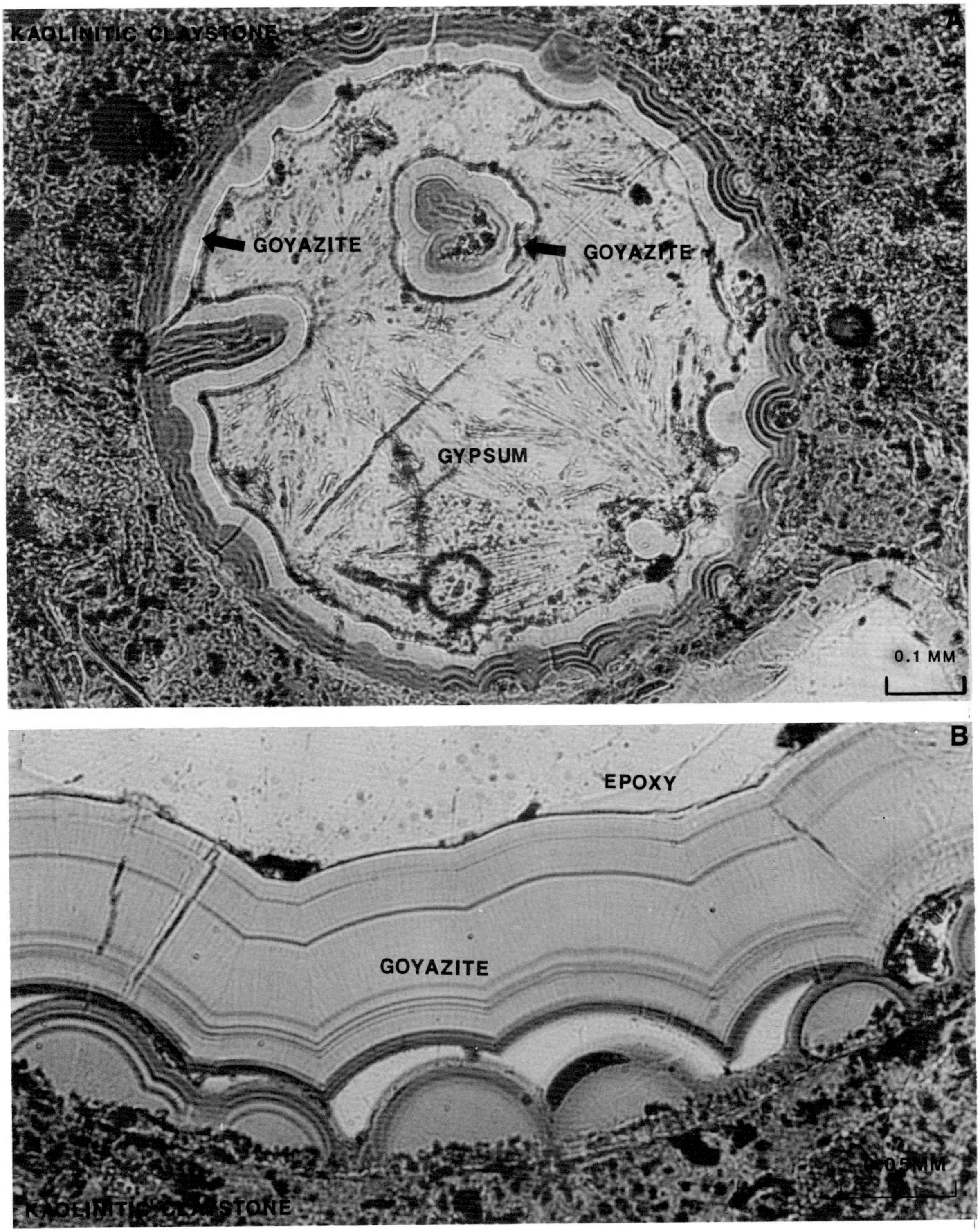

Figure 21. Photographs of thin sections containing goyazite spherules from the Dogie Creek, Wyoming, K/T boundary locality. A, shell of spherule composed of colloform microlaminated goyazite. Center of spherule filled with gypsum. Goyazite inclusion(?) in upper part of spherule. B, details of the structure of the shell of a goyazite spherule. Colloform, fibrous goyazite filled a spherical cavity in the kaolinitic claystone of the K/T boundary claystone.

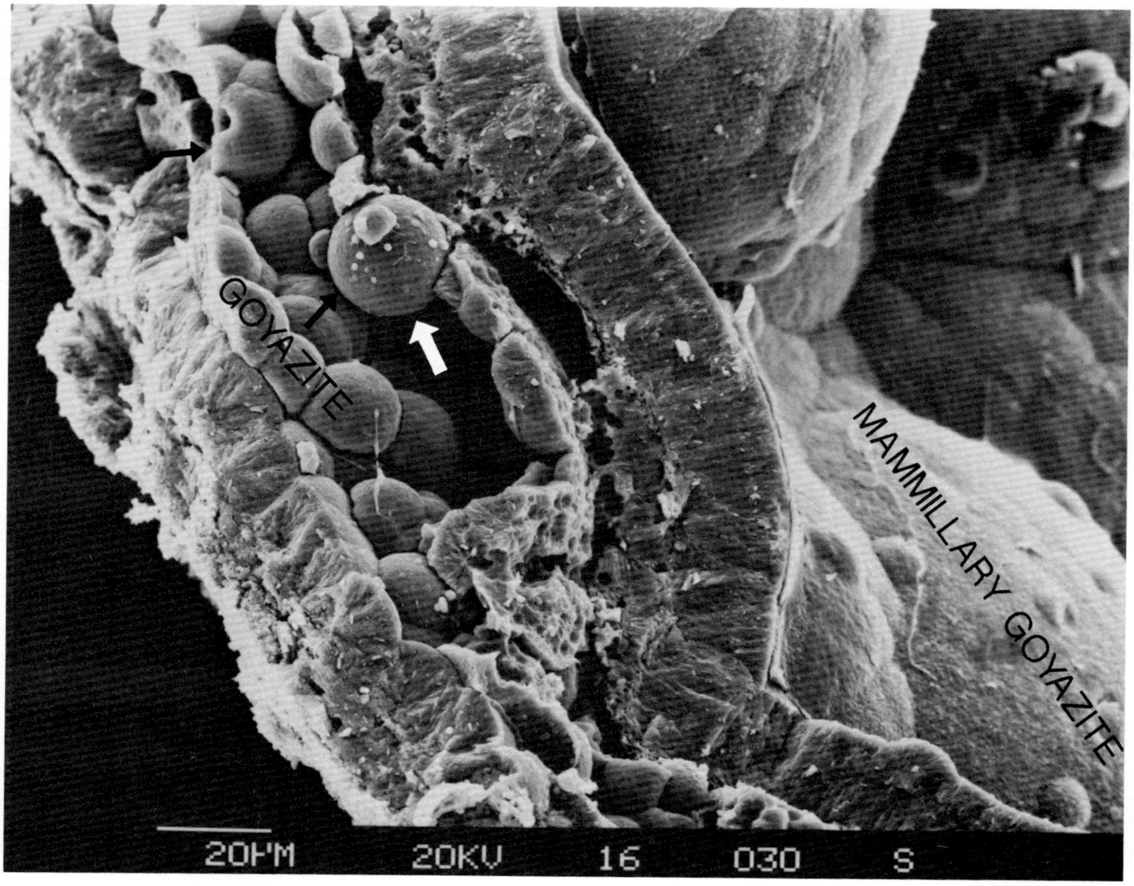

Figure 22. Shell of goyazite spherule from the K/T boundary claystone at the Teapot Dome, Wyoming, locality. The shell of the spherule is composed of two layers of microlaminated goyazite. Note vug filled with goyazite spherules (shown at arrows).

polarized light. EDS-SEM study of the darker laminae of the rims or shells indicated that they are richer in iron than adjacent laminae. The shells or rims of some goyazite spherules are compound; that is, they consist of two colloform layers of goyazite and an intervening void filled with mammillary and microspherulitic goyazite (Fig. 22).

Many of the large goyazite spherules are hollow, some are filled or partially filled with granular or bladed gypsum, and a few are partially filled with either kaolinite, barite, marcasite, and jarosite or a mixture of these. Thin sections show that the hollow goyazite spherules were plastically deformed (Fig. 18B) or crushed (Fig. 20A); however, spherules filled with gypsum were, for the most part, not deformed or crushed. Among the crushed spherules, broken shell fragments are surrounded by gypsum. Crushed and deformed hollow spherules are flattened parallel to bedding and occur adjacent to uncrushed ones. Lenses, stringers, and irregularly shaped masses of the broken walls of spherules occur among uncrushed spherules. Some goyazite spherules have circular depressions or dimples on their surface, and interestingly, these dimples are also recorded on successively deeper goyazite shells (Fig. 19A, C). These dimples probably formed by compaction of adjacent goyazite spherules during an early stage of diagenesis when the goyazite spherules were soft.

Numerous goyazite spherules are compound; that is, they appear in section to consist of a series of nested, thin-walled spherules. Each shell of the series is smaller in diameter than its outboard neighbor, and each shell is separated from its neighbor by a spherical shell of gypsum. One peculiar goyazite spherule consisted of two thin-walled oblate spheroid shells separated by gypsum and a central, thin-walled, spherical one filled with gypsum.

The combined weight of the observational evidence described in the foregoing paragraphs suggests to me that (1) several growth generations of goyazite spherules occur in the layer, and (2) the goyazite mineralization occurred early in the diagenetic history of K/T boundary sediments.

Origin of spherules

The origin of the spherules in the K/T boundary claystone in the Western Interior of North America is a controversial subject. Smit (1984), Pollastro and others (1983), Smit and Romein (1985), Bohor and others (1987), Pollastro and Pillmore (1987), and Pillmore and Flores (1987) have either described or noted the occurrence of K/T boundary spherules. Two different modes of impact origin have been suggested for the kaolinite and goyazite spherules. One suggestion has been that the goyazite spherules were originally microtektites; that is, they were originally solid glass spherules that formed from crustal target rocks during the earliest phase of the K/T impact event. Smit (1984) and Smit and Romein (1985) referred to them as being "microtektite-like" because of their spherical shape and their occurrence in a bed thought by them to consist of altered impact ejecta. Although they referred to the spherules as being microtektite-like, the clear implication was that the spherules were considered to be altered microtektites.

A second suggestion is that the goyazite and kaolinite spherules formed during the K/T boundary impact as hollow glass bubbles or spherules. Pollastro and Pillmore (1987) maintained that the kaolinite spherules were originally glass-walled bubbles, and these altered to kaolinite along with fragmental glassy material that now is their host, the boundary claystone. They claimed that the bubble walls altered from glass to columnar kaolinite through an amorphous gel stage, and the bubble cavities were filled subsequently with microspherulitic kaolinite similar to that of the boundary claystone. They did not explain why the glassy bubble walls altered to columnar kaolinite whereas the fragmental glass of the bubble host altered to microspherulitic kaolinite. They also did not explain by what process the bubble cavities were filled with microspherulitic kaolinite.

Bohor and Triplehorn (1987) and Bohor and others (1987, p. 899) also speculated that the goyazite spherules formed during the K/T boundary impact as hollow glass spherules. They imagined that the spherules were derived from the ". . . depressurized rocks exposed at the impact-crater floor." Strangely, these authors did not speculate as to the origin of the kaolinite spherules.

The kaolinite and goyazite spherules in the K/T boundary claystone in the Raton basin were thought by some (Smit and Romein, 1985) to be analogous to particles (described as spherules) composed of sanidine discovered by Smit and Klaver (1981) in marine K/T boundary rocks at Barranco del Gredero, Caravaca, Spain. They surmised that the sanidine particles in Spain formed by solidification of melt droplets and were derived from the impacting body. Montanari and others (1984) found particles of potassium feldspar (sanidine of Smit and Klaver, 1981) and mixtures of glauconite, magnetite, quartz, and pyrite at K/T boundary sites in northern Italy and in a deep-sea core from hole 465A in the central Pacific Ocean (see also Smit and Kyte, 1984; Kyte and Smit, 1986). Montanari and others (1984) studied the potassium feldspar particles and, based on their internal texture, claimed that they formed diagenetically from calcic plagioclase melt droplets. They speculated that the droplets formed by condensation of vaporized rock in the high-temperature fireball and cloud produced by the K/T impact.

Interpretation by Shaw and Wasserburg (1982, p. 175) of isotopic data for the potassium feldspar particles led them to suggest that they formed as a result of ". . . an oceanic impact in which both basaltic oceanic crust and overlying sediments or seawater were involved." They indicated that the isotopic results are compatible with a volcanic origin (authigenic growth from volcanic detritus) for the potassium feldspar. However, a continental source for the progenitor of the sanidine was ruled out by them. They concluded that a major problem still exists because it is difficult to understand how to make potassium-rich sanidine that is extremely poor in strontium from any reasonable oceanic crust progenitor by an impact process.

DePaolo and others (1983) determined isotopic ratios and trace-element compositions of K/T boundary rock units at Caravaca, Spain. They found that the strontium, neodymium, oxygen, and hydrogen isotopic composition of the thin reddish brown layer at the K/T boundary differed from adjacent Cretaceous and Tertiary marl. They (p. 370) concluded that the present chemical composition of the potassium feldspar of the so-called "sanidine spherules" in the impact layer ". . . is attributable to authigenic growth." They further concluded that the isotopic composition, low REE content, and high concentration of nickel in this layer are most easily explained if the K/T impact occurred in oceanic crust and if a substantial fraction of the impact target material were from relatively deep in the lithosphere.

Hansen and others (1986) studied potassium feldspar and iron-oxide particles in marine K/T boundary rocks in Europe, and have judged that their morphology and structure do not demonstrate an impact origin. Instead, the particles are the diagenetic infill of marine green algae (Prasinophyta). According to these authors (1986, p. 870), the algae are a disaster species; their numbers increase explosively if the normal plankton population markedly declines.

Izett (1987a) also studied K/T boundary samples from Caravaca, Spain, that contain the sanidine particles referred to by some as spherules, and he inferred they may not be melt droplets of impact derivation. The potassium feldspar particles are probably not altered basaltic melt droplets because their shapes are highly irregular and not compatible with the shape expected of melt droplets. For example, sanidine particles in the impact layer at Agost and Caravaca, Spain, are highly variable in shape. Spherical and highly irregular-shaped particles occur within a fraction of a millimeter of each other, and the irregular shapes of some, as opposed to the subspherical shape of adjacent particles, cannot be selectively attributed to deformation of the enclosing sediments. The open-meshwork texture of the potassium feldspar particles also suggests that they did not form by diagenetic alteration of melt droplets having an igneous quench texture as proposed by Montanari and others (1984). Rather, the texture of the crushed and uncrushed sanidine spherules is authigenic.

Moreover, the sanidine particles occur in claystone units stratigraphically below and above the impact layer at Caravaca. In this regard, Rosenfeld and others (1989) reported finding spherules, some of which are potassium feldspar, in three different layers in upper Maastrichtian marine rocks of southern Israel.

Recently, F. E. Lichte and I made LA-ICP-MS analyses of hand-picked sanidine spherules from the Caravaca, Spain, site. The amounts of about 75 elements, in the parts per billion range, were measured in one series of laser pulses. Of the elements measured, the content of the platinum-group metals was of special interest. The analyses showed that the sanidine spherules contain less than 8 ppb iridium and other platinum-group metals, whereas the host impact-layer claystone contains as much as 57 ppb iridium (Kyte and others, 1985). The nickel content of the sanidine also was of special interest because several geologists believe that the nickel in K/T boundary rocks is of extraterrestrial impact origin (Kyte and others, 1985; Gilmour and Anders, 1989). The sanidine contains only about 480 ppm nickel, whereas the claystone that surrounds the sanidine spherules contains an order of magnitude more (Kyte and others, 1985). The nickel probably is located in hydrous iron-oxide inclusions (formerly pyrite or marcasite) in the sanidine spherules (Izett, 1987a; Fig. 7A). The foregoing chemical evidence suggests, but does not prove, that the sanidine spherules are not altered impact glass. If the sanidine spherules originally were altered impact glass, it logically should contain as much nickel and platinum-group metals as the host impact-layer claystone, also presumed to be altered impact glass.

In conclusion, some geologists believe that the goyazite and kaolinite spherules of the Raton basin are of impact origin because they occur in a bed, the boundary layer claystone, that they also believe is of impact origin. Be this as it may, the following information seems to be inconsistent with the idea that the goyazite and kaolinite spherules were originally microtektites of impact origin:

1. Spherules only occur in the K/T boundary claystone and not in the overlying K/T boundary impact layer, which is the chief stratigraphic location of an iridium abundance anomaly and shock-metamorphosed minerals. Seemingly, the spherules (alleged microtektites) and the iridium anomaly should occur in the same layer, composed in part of early time, high-velocity impact ejecta.

2. Spherules are common (as much as 30 percent) at two Wyoming K/T boundary sites, but nonexistent at a third nearby (about 45 km) site. Spherules are rare (far less than 1 percent) at Raton basin sites, and absent at Canadian (Alberta and Saskatchewan) sites. Such a variable abundance for the alleged microtektites, although not impossible, is difficult to reconcile with an impact origin.

3. Some spherules are microtektite-like in their external shape, but a few are not spherical and form peculiarly shaped objects (Izett, 1987a, Fig. 9) unlike microtektites illustrated by Glass (1984, Fig. 5). For example, no summer squash– or cigar-shaped objects such as those in the boundary claystone (Fig. 19) have been reported in assemblages of microtektites.

4. Many goyazite spherules are hollow and unlike most microtektites, which are solid (see Glass, 1984).

5. The surfaces of many of the goyazite spherules are smooth and unlike the pitted and corroded surfaces of most tektites and some microtektites (see Glass, 1984, p. 338).

6. If some of the spherules were originally solid microtektites that fell into standing water, then because of their size and density, they seemingly would be concentrated at the base of the claystone bed. However, they are not concentrated at its base, but are widely distributed throughout the bed in the Raton basin.

7. The coexistence of spherules of markedly different compositions (kaolinite and goyazite) within a millimeter of each other makes an impact origin for them unlikely. If the original material for each of these was a compositionally uniform impact glass, it is difficult to understand how one glassy spherule would alter to goyazite and an adjacent one alter to kaolinite.

8. In the Raton basin, kaolinite spherules are significantly larger than goyazite spherules. If they were formed from the same population of microtektites, it is remarkable that the large ones would alter to kaolinite and the small ones to goyazite.

9. In the Raton basin, kaolinite and goyazite spherules are widely disseminated in the boundary claystone, but locally they occur in compact clusters. This clustering distribution is not consistent with the idea that the boundary claystone is an air-fall or a reworked air-fall unit.

10. The occurrence of aluminum phosphate (goyazite) spherules in the K/T boundary claystone is not unique; aluminum phosphate spherules also occur in a tonstein of volcanic origin in Kentucky and a coal bed in Wyoming.

11. The low concentration of nickel (4.2 ppm) and platinum-group elements ($<$ 8 ppb) in the goyazite spherules suggests, but does not prove, that they are not altered tektite glass.

In addition to the foregoing evidence, the following information also seems to be inconsistent with the idea that the goyazite and kaolinite spherules were originally hollow glass spherules of impact origin.

1. The walls of the goyazite spherules are compound and composed of microlaminated colloform goyazite and not simply the replaced glass shells of spherules of impact origin.

2. The walls of goyazite spherules are locally vuggy, and the vugs are filled with goyazite spherules. These goyazite spherules are similar but smaller than those alleged to be altered microtektites.

If the kaolinite and goyazite spherules are not of impact origin, what then is the origin of these enigmatic objects? A complete answer to this question is not known, but what is known is summarized as follows. At Raton basin K/T boundary sites, spherical cavities of unknown origin (dissolution of microtektites, soft organic objects, or methane gas bubbles?) formed within the progenitor sediment of the boundary claystone. The cavities were filled with either a thin shell of columnar kaolinite, a thick shell of goyazite, or completely by goyazite. The central parts of those spherical cavities that were lined with a thin shell of columnar kaolinite were filled with microspherulitic kaolinite, whereas the central parts of those cavities lined with a thin shell of goyazite were filled with platy kaolinite. Kaolinite spherules only formed at sites where the boundary claystone is composed of kaolinite, not smectite. At Wyoming K/T boundary sites, spherical and other odd-shaped cavities were filled generally with a thin shell of goyazite. Subsequently, the cavities were filled with gypsum, kaolinite, barite, jarosite, and marcasite or a mixture of these. Vugs formed within some goyazite shells, and goyazite spherules grew within these vugs. Compaction of the sediments and crushing of the goyazite spherules directly followed the goyazite-mineralization event.

Origin of the K/T boundary claystone

Although the K/T boundary claystone of the Raton basin is a useful geologic mapping tool and a stratigraphic guide for identifying and locating the K/T lithologic and paleontologic boundaries, its origin is controversial and an enigma. Ideas have been advanced that the K/T boundary claystone is a primary air-fall unit of impact or terrestrial volcanic origin. However, some information, including microstratigraphic relations, lithologic and petrographic observations, and chemical data, accumulated during my study are not in accord with either an impact or volcanic origin for the K/T boundary claystone.

Alvarez and others (1980) proposed that the K/T boundary claystone in marine rocks of Europe consists of "pulverized rock" of impact origin. If they were correct that the boundary claystone was originally composed of pulverized, shock-metamorphosed target rocks, the distal fallout material probably should contain some clastic mineral grains that almost surely would have survived diagenesis. This conclusion is based on the fact that quartz and feldspar grains less than 100 μm in diameter in the impact layer are not altered. Accordingly, if the boundary claystone were composed of impact-derived material, it should contain a considerable amount of shocked quartz and feldspar; as will be pointed out in a subsequent section of this report, however, it generally does not.

The conclusion that the K/T boundary claystone in the Raton basin is of impact origin was based on two discoveries: (1) that an iridium anomaly occurs at the boundary (Orth and others, 1981), and (2) that a thin claystone layer invariably marks the boundary (Pillmore and Tschudy, 1983; Pillmore and others, 1984). The idea that this claystone layer is of impact origin was supported further by the discovery that boundary-interval rocks contain shocked quartz grains (Izett and Pillmore, 1985a). Pollastro and Pillmore (1987) emphasized that the boundary claystone (1) forms a thin, discrete unit, whose geometry is compatible with the type of unit they visualized for a distal air-fall impact ejecta layer; (2) constitutes a widespread stratigraphic marker that occurs sporadically from New Mexico to near Edmonton, Canada (Izett, 1987c, 1988); and (3) contains above-background amounts of platinum-group elements such as iridium.

A second idea is that the K/T boundary claystone of the Raton basin is of volcanic origin. Alvarez (1983) proposed that the boundary claystone in marine rocks formed, at least in part, from the alteration of a volcanic ash. I presume that he had in mind a normal silicic volcanic ash as judged by its regional extent. If this were true, the ash, as erupted, should have contained components of an assemblage of primary volcanic minerals typical of voluminous, widely dispersed, volcanic ash beds. On the contrary, it does not. The boundary claystone in the Raton basin and elsewhere in western North America characteristically lacks clastic mineral grains (except where it was reworked or bioturbated) and, in particular, generally lacks crystals or broken crystals of quartz and feldspar of volcanic origin. Study of hundreds of volcanic ash beds has shown that most regional silicic ash beds commonly contain quartz, feldspar, zircon, apatite, or ilmenite phenocrysts of *uniform chemical composition* (Izett, 1981). It would be a rare volcanic ash that did not contain one of these stable mineral species as a primary phase. Surely, some of these minerals would survive diagenesis of ash to kaolinite (Ryer and others, 1980; Spears and Rice, 1973). Moreover, many Cretaceous bentonites of the Western Interior of North America contain biotite, quartz, feldspar, and zircon. Nevertheless, a volcanic origin for the K/T boundary claystone cannot be rejected dogmatically, although its apparent lack of volcanic phenocrysts seems to rule out this origin.

Pillmore and others (1984) apparently agreed with the conclusion of Alvarez (1983) that the boundary claystone is, in part, of volcanic origin. They contended that in the Raton basin it formed by a combination of volcanic and impact processes. This idea was based seemingly on the observation that boundary claystone is similar to tonsteins and cryptotonsteins in coal beds presumably of volcanic origin. Some geologists, including Spears and Rice (1973) and Bohor and Pillmore (1976), have maintained that tonstein layers are altered volcanic ash, and indeed, many may be. Others have favored an origin involving either the biogenic accumulation of a silica-alumina gel or the accumulation and alteration of a previously formed clay. Williamson (1970) presented a summary of the various theories concerning the origin of tonstein layers. Seiders (1965) demonstrated that a thin flint clay or tonstein in the Breathitt Formation (Pennsylvanian age) of eastern Kentucky is of volcanic origin on the basis of its content of sanidine and beta quartz crystals. My study of the insoluble residue of this tonstein, which includes euhedral zircon, beta quartz, and sanidine, also indicates unquestionably that this tonstein is of volcanic origin. For example, the zircon crystals are euhedral, have similar morphologies, and all have about the same color. Thus, they form a *coherent assemblage.* It is interesting to note that if an iridium anomaly and shock-metamorphosed minerals had not been found associated with the boundary claystone tonstein, a volcanic origin for it would be assigned without question by some geologists.

The trace-element chemical composition of the boundary claystone also is incompatible with the idea that it consists of altered silicic volcanic ash, as pointed out by Gilmore and others (1984). Moreover, the low TiO_2 to Al_2O_3 ratio (0.047) of nu-

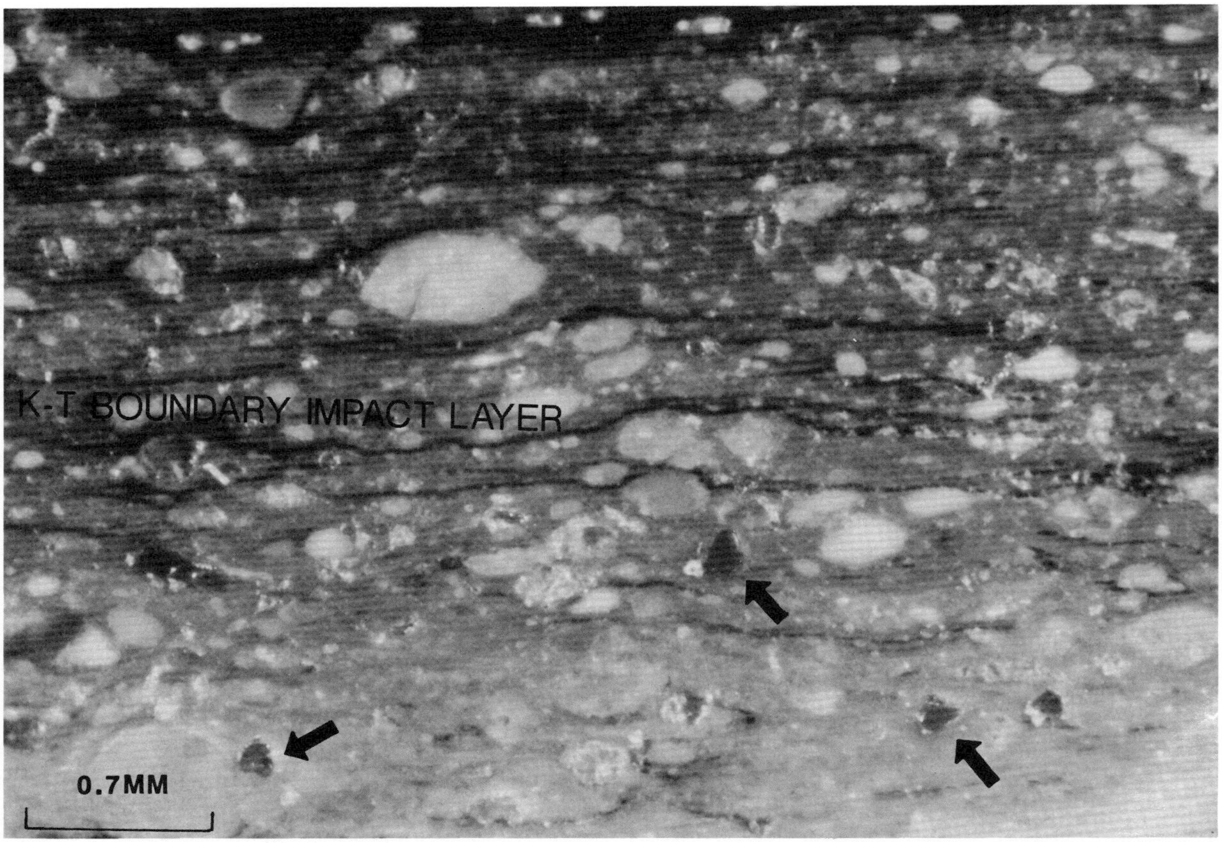

Figure 23. Polished surface of a slab of the K/T boundary impact layer at the Berwind Canyon K/T boundary site about 20 km northwest of Trinidad, Colorado. Impact layer typically contains laminae of black vitrinite and white almond-shaped kaolinite pellets. Scattered small clastic quartz grains shown at arrows. Some quartz grains contain shock lamellae. Photographed with incident light.

merous analyzed samples of the claystone (Table 2) indicates that, if the precursor material were volcanic glass, it could not have been of basaltic composition (compare Spears and Kanaris-Sotiriou, 1979, Fig. 4). Basaltic ash is the only conceivable type of volcanic material that could contain iridium in amounts comparable to those in K/T boundary rocks, and even this possibility seems remote because basaltic material lacks sufficient amounts of platinum-group elements to account for the magnitude of the K/T boundary iridium anomaly (see Strong and others, 1987, p. 2775).

Information supporting the contention that the boundary claystone is distal air-fall material from either a volcanic or impact event is based mainly on two observations: its geometry and its texture. Concerning the first observation, the geometry of the bed, Gilmore and others (1984, p. 228) maintained "The uniformity in thickness of the K/T kaolinitic clay bed over a distance of 1,200 km indicates that the fallout material came from a common source at a considerable distance." This evidence and their argument are persuasive, but the geometry of the bed does not prove conclusively that it formed from air-fall material of either impact or volcanic origin. Morever, it should be emphasized that the boundary claystone does not form a continuous blanket-like layer in K/T boundary rocks of western North America as Gilmore and others (1984) imagined. The boundary claystone has been found only at spot sites, generally in a fine-grained sequence of rocks that were deposited in or adjacent to coal swamps.

Concerning the second observation, Pollastro and Pillmore (1987) referred to the texture of the claystone as "imbricate" and claimed it was inherited from a very coarse-grained fragmental impact-fallout material consisting of vesicular glass shards. Although, superficially, the texture resembles that of pyroclastic rocks, it lacks angular shard shapes (see Fig. 4B) and, more important, lacks broken crystals typical of these rocks (Ross and Smith, 1961, Figs. 16–23).

Information not in accord with an impact or volcanic origin for the K/T boundary claystone consists of stratigraphic relations, lithologic observations, and chemical data. Each of these is discussed in the following paragraphs.

A stratigraphic relation of considerable interest was discovered by me during study of Raton basin K/T boundary rocks. I found that large, shock-metamorphosed mineral grains are concentrated (Fig. 23) in a thin claystone layer (herein called the K/T impact layer) that overlies the K/T boundary claystone bed (Izett and Pillmore, 1985a, 1985b; Izett, 1987c, 1988). This information suggests that the boundary claystone may not be composed of impact material. If it were, it is difficult to envision a single, air-fall-depositional event that would concentrate large, shock-metamorphosed mineral grains in the upper rather than the lower unit of the boundary interval couplet. Stokes' law predicts that large grains should be concentrated at the base of the couplet if it were composed of primary air-fall material. Moreover, it is difficult to understand how the K/T boundary claystone could originate from an impact glass free of shock-metamorphosed minerals. Stöffler (1984, p. 466) and Grieve (1987, p. 253) pointed out that an important feature of impact glasses is their content of shocked mineral grains and lithic clasts.

A second stratigraphic relation that suggests a nonimpact origin for the boundary claystone of the Raton basin is that an exact equivalent of the boundary claystone is apparently absent in marine K/T boundary rocks in Europe. At nonmarine depositional sites in western North America, the impact layer of this paper records the abrupt appearance of shock-metamorphosed minerals. At marine depositional sites such as Caravaca and Agost, Spain, and Stevns Klint, Denmark, shock-metamorphosed minerals first appear and are concentrated in a 1- to 3-mm-thick ferruginous clay unit that rests on Cretaceous marl and marks the K/T boundary (Izett, 1987a). This ferruginous clay unit also contains an iridium concentration spike and marks the horizon of a significant mass-extinction event (Smit and Hertogen, 1980). Thus, there is no equivalent of the Western Interior North American K/T boundary claystone below the thin ferruginous impact layer and above Cretaceous marl at Caravaca and Agost, Spain, or Stevns Klint, Denmark (Fig 24; see also back cover of this paper). Therefore, the impact layers of western North America and Europe record the first arrival of shocked minerals in the boundary interval. Then, seemingly, the boundary claystone, which underlies the impact layer in western North America, but not in Europe, may not be of impact origin. If it were of impact origin, one would expect to find its equivalent in European K/T boundary sections.

A third stratigraphic relationship bears on the origin of the K/T boundary claystone. At most sites in the Raton basin, the boundary claystone is sharply overlain by the impact layer (Fig. 1). The contact between these units separates dissimilar lithologies and marks a subtle change in depositional style. At some places, however, the contact appears gradational, but only within a 1-mm interval. Within this interval the claystone appears to be turbated and degraded into proto-pellets (Fig. 11A). Thus, the evidence seems to suggest that the top of the boundary claystone was a paleosurface upon which plants grew (Figs. 1 and 7) and the impact layer was deposited.

A fourth stratigraphic relation suggests that the boundary claystone is not composed of air-fall material. Typically, the contact between the boundary claystone and underlying carbonaceous shale or mudstone is gradational in detail. This relation is at odds with the nature of the lower contacts of air-fall units. Characteristically, air-fall volcanic ash beds have sharp contacts with underlying sediments.

Two lithologic observations are not in accord with an impact origin for the boundary claystone. First, the boundary claystone consists of relatively homogeneous claystone containing accordion-like carbonaceous root-like structures and deformed stringers of carbonaceous material; with the enclosing claystone, it has a texture resembling a marble cake. Furthermore, the impact layer consists of laminated claystone that contains planar vitrinite laminae. This information implies that the boundary claystone was deposited and its included carbonaceous material was deformed prior to deposition of the overlying layer that records the K/T impact event.

A second lithologic observation was made at three sites in the Raton basin (Clear Creek North, Carmel, and Madrid Railroad). At these sites, the boundary claystone is missing and the impact layer rests directly on carbonaceous shale (see Fig. 10), indicating that the boundary claystone was not deposited or eroded prior to deposition of the impact layer.

Other lithologic information also is not consistent with an air-fall impact origin for the boundary claystone. Apparently the boundary claystone accumulated slowly by normal fluvial or lacustrine sedimentary processes rather than rapidly by atmospheric fallout of impact material. Its content of kaolinite pellets, centimeter-long horizontal vitrinite laminae, blebs of amber, degraded plant material, fossil twigs, rounded fragments of woody material (Fig. 8), and carbonaceous sedimentary structures (probably roots) that flare out at the contact wit the impact layer (Fig. 7B) all argue for a complex origin for the boundary claystone. At several K/T boundary sites the boundary claystone contains a considerable amount of unshocked quartz grains incorporated into the claystone during fluvial sedimentation events.

Gilmore and others (1984) and Pollastro and Pillmore (1987) thought that the claystone in the Raton basin formed by alteration of glassy impact ejecta similar to tektite glass. However, there is indirect evidence that the K/T boundary claystone could not be altered tektite glass. A fact apparently not appreciated by many geologists is that tektite glass is markedly different than ordinary volcanic glass. It is essentially anhydrous and has a relatively stable structure. Adams (1984, p. 202) stated, "The extremely low water content of tektites makes them virtually impregnable to hydration." For example, tektites and microtektites of upper Eocene age have survived in pristine condition since 35.4 ± 0.6 Ma (Obradovich and others, 1989) in sediments of DSDP site 612 on the continental slope off New Jersey (Glass, 1987; Thein, 1987). It is plausible that some tektite glass that formed during the K/T event would still be glass, and would not have altered to kaolinite as some have suggested (Pillmore and others, 1984; Pollastro and Pillmore, 1987). In contrast, volcanic glasses (as erupted) contain a relatively large amount, a few tenths of 1 percent, of volatiles (see Ross and Smith, 1955). Because tektite glass does not hydrate under ambient conditions, it does not alter to clay or zeolites as does volcanic glass. Instead, in an

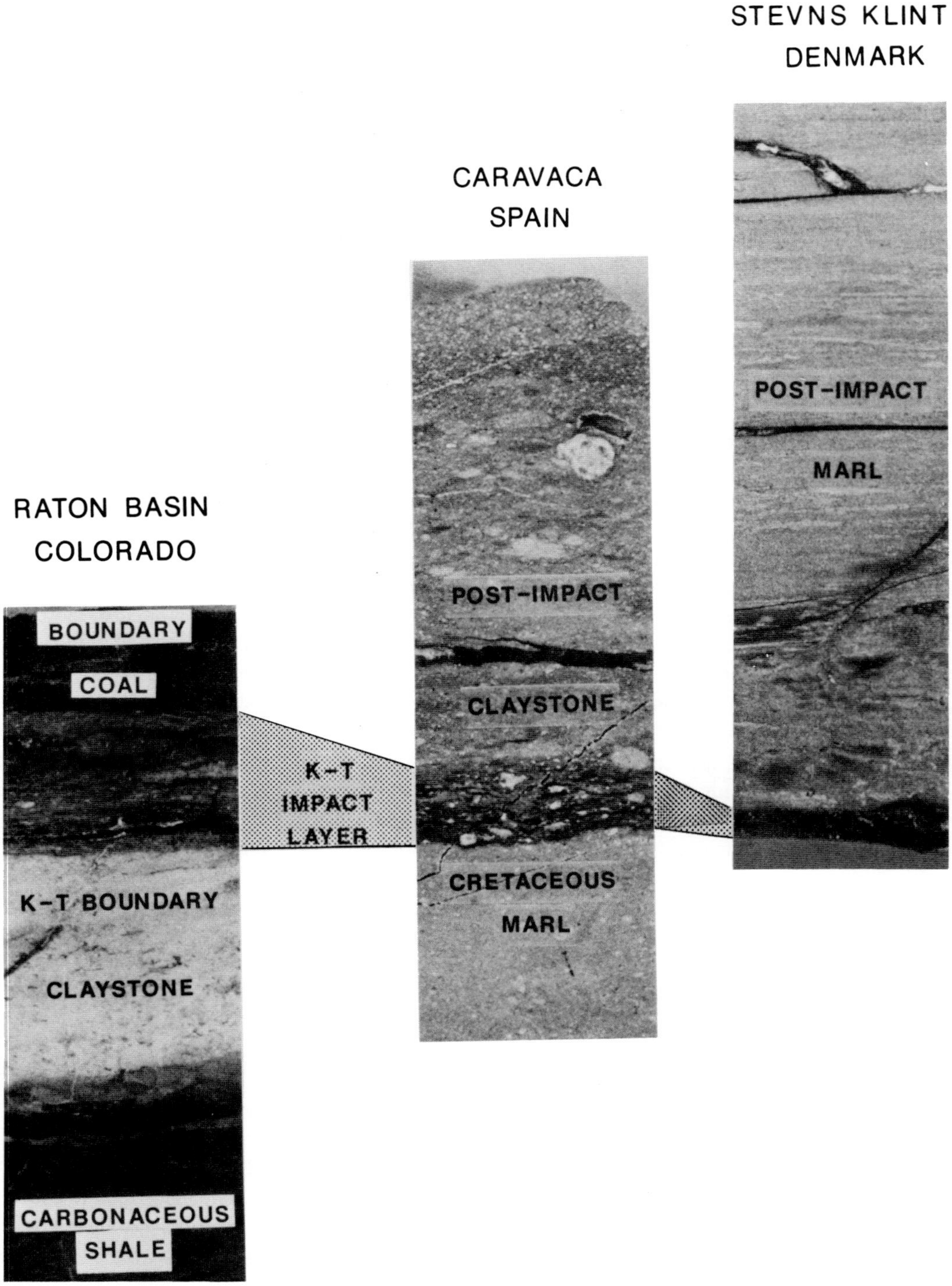

Figure 24. Comparison of K/T boundary interval rocks deposited in continental and marine environments from North America and Europe. Polished slab of continental K/T boundary rocks from the Clear Creek North K/T boundary site, about 10 km south of Trinidad, Colorado. Impact layer is 6.0 mm thick. K/T boundary interval marine rocks from the Caravaca, Spain, locality. Impact layer, 1.0 mm thick, is underlain by Cretaceous marl. K/T boundary interval marine rocks from the Stevns Klint, Denmark, locality. Impact layer is 1 to 3 mm thick. Marl underlies the impact layer, but it was lost in preparing the sample. The impact layer records the abrupt appearance of shocked minerals and an iridium abundance anomaly. No equivalent of the K/T boundary claystone bed of the Western Interior of North America occurs at marine K/T boundary sites in Europe.

aqueous environment, tektite glass dissolves, and accordingly, tektites have corroded, rough surfaces. If this process went undisturbed to completion, the tektite glass would be completely dissolved.

If the claystone bed contained some altered asteroidal impact ejecta, it seemingly should contain a significant iridium anomaly as does the overlying impact layer. It quite simply does not. Data of Gilmore and others (1984, p. 226), Pillmore and others (1984), Pillmore and Flores (1987), and data of this paper illustrate the fact that amounts of iridium are generally highest in the impact layer and adjacent carbonaceous-rich rocks and lowest in the boundary claystone.

The exact origin of the claystone remains a paradox, but some of its microscopic textural features and chemical features suggest that it has a complicated diagenetic history. It is by definition a tonstein, but one that is different texturally and slightly different chemically from other tonsteins in the Raton basin.

In summation, the contention that the boundary claystone is of impact origin is based mainly on two observations: (1) it forms a thin, widespread bed in continental sedimentary rocks; and (2) it forms a stratigraphic couplet with the overlying K/T boundary impact layer at nearly all K/T boundary sites that contain anomalous amounts of iridium. The first observation, that the boundary claystone is a thin, widespread unit, is well documented, but that is not to say it formed as a continuous air-fall, blanket-like layer. It is important to remember that the thin beds of the boundary interval (boundary claystone, impact layer, and boundary coal bed) have been found at only about 30 sites in the Western Interior of North America, and the rocks at each of these sites formed in low-energy, coal-swamp–related environments that existed in a series of north-trending Rocky Mountain coal basins. The boundary claystone has not been found outside these coal basins. The second observation, which suggests that the boundary claystone is of impact origin, is based on the stratigraphic association at nearly all sites of the boundary claystone and impact layer. This may be considered an argument for a genetic relation between the two units. However, it should be remembered that the impact layer is overlain at most sites in Western Interior North America by a thin coal bed, which certainly is not of impact origin.

The most important facts that bear on the origin of the K/T boundary claystone are that (1) the boundary claystone forms a distinctive sedimentary unit lying below the K/T boundary impact layer, and (2) shock-metamorphosed minerals are concentrated in the impact layer and not in the underlying boundary claystone. As pointed out previously, it is difficult to imagine how the boundary claystone could originate from impact material that would be essentially free of clastic mineral grains, in particular, shock-metamorphosed minerals.

If the K/T boundary claystone is not of impact origin, as indicated by some observational evidence, reasonable questions are (1) what is its origin? and (2) how could such a thin bed of claystone form over such a broad area by normal sedimentary processes? The answers to these questions are not known, but the boundary claystone has several features suggesting that its upper boundary is a paleosurface and that the claystone underwent pedogenic processes. The claystone should be thoroughly studied by soil scientists familiar with microscopic characteristics of pedogenic claystone formed in tropical soil-forming environments (Retallack, 1976; Fastovsky and others, 1989). To form a claystone over such a broad area would require amazingly uniform depositional conditions, and indeed such conditions may have existed during the Cretaceous. Kauffman (1987, p. 150) pointed out that climatic conditions at the close of the Cretaceous were much more equable than at present and were characterized by lack of polar ice, low seasonality, and broad tropical-temperate climate zones.

K/T BOUNDARY IMPACT LAYER

A thin microstratigraphic unit of claystone overlies the K/T boundary claystone in the Raton basin (Fig. 1). The most surprising and puzzling observation I made in late 1984 during study of K/T boundary rocks is that this unit, rather than the underlying boundary claystone, contains a concentration of shock-metamorphosed mineral grains (Izett and Pillmore, 1985a, 1985b; Izett and Bohor, 1986a). This observation has important implications concerning dispersal-deposition mechanics and arrival time of impact ejecta in the K/T boundary interval. The precise stratigraphic location of the shock-metamorphosed materials was determined by microscopic study of about 100 polished specimens of K/T boundary interval rocks similar to the polished slabs shown in Figure 1. The exact location of quartz grains was revealed by polishing ground surfaces of the claystone with diamond paste as described previously. Owing to their greater hardness, the quartz grains stand in conspicuous relief above the softer surrounding claystone. As seen on polished surfaces, the quartz and other mineral grains are set in the claystone like raisins in a pudding (Figs. 11A and 23). Study of numerous thin sections of the impact layer suggests that there might be a slight tendency for the largest grains to be concentrated at the base of the unit.

It should be emphasized that the magnitude of the shock-metamorphic mineral anomaly or spike is enormous because rocks below the K/T boundary interval are virtually devoid of shocked minerals. This spike of shocked mineral grains occurs at each of the 20 K/T boundary sites in the Raton basin in the same microstratigraphic sequence and unit.

The impact layer in the Raton basin is composed mainly of kaolinite and I/S mixed-layer clay containing as much as 50 percent illite layers, according to C. G. Whitney (U.S. Geological Survey, written communication, 1986). The claystone differs from that of the boundary claystone as shown by x-ray diffractograms of glycolated samples heated to 550°C. For an unknown reason, the structure of the kaolinite and I/S mixed-layer clay of the boundary impact layer collapses and forms a sharp peak on diffractograms at about 8.8° 2*θ*. In contrast, the structure of the kaolinite and mixed-layer clay of the boundary claystone collapses in the same region to nearly a flat trace.

Microscopic study of mineral concentrates obtained from the impact layer from sites in the Raton basin and elsewhere in western North America showed that they consisted of (1) single grains of quartz, chert, chalcedony, oligoclase, and potassium feldspar; and (2) composite grains (lithic fragments) composed of metaquartzite, quartzite, and granite-like aggregates of quartz, oligoclase, and potassium feldspar (microcline). In addition, mineral concentrates contain a miniscule amount of other stable mineral grains (Izett and Pillmore, 1985a; Izett, 1987c). Most mineral concentrates contain rounded grains of zircon (some metamict) and tourmaline; some contain a few grains of allanite, garnet, epidote, chromium spinel, ilmenite, and metamorphic blue-green hornblende; others contain extremely rare grains of beta quartz, orthoclase, and sanidine. Authigenic minerals in the impact layer from western North American sites include rutile, barite, gypsum, pyrite, marcasite, jarosite, and hematite. The presence of extremely rare grains of volcanic minerals such as beta quartz and sanidine in the impact layer at a few sites does not indicate that the claystone is of volcanic origin because bentonites containing these minerals are common in Cretaceous rocks of the Western Interior of North America. Similarly, a few grains of volcanic minerals in a Cretaceous sandstone certainly would not indicate it had a volcanic origin.

A similar suite of detrital heavy minerals occurs in sedimentary rocks stratigraphically below and above the impact layer. This information, combined with the observation that most of these minerals do not occur in the impact layer at marine K/T boundary sites, implies that they are not atmospherically transported impact target materials. Presumably, these minerals were incorporated into the impact layer during its deposition and were derived from the erosion of older sedimentary rocks and Precambrian metamorphic and plutonic rocks.

Exceptionally rare, minute (20 to 40 μm), perfectly formed crystals of a spinel (magnesioferrite) occur in K/T boundary marine sediments in Europe (Bohor and others, 1986; Kyte and Smit, 1986). Such crystals are thought by these authors to have formed during the K/T boundary impact event. Izett (1987a) also studied magnesioferrite crystals in the K/T boundary impact layer at Caravaca, Spain, and pointed out several problems with an impact origin for the crystals. Large samples of the K/T boundary impact layer were collected from several sites in the Raton basin with the purpose of searching for magnesioferrite crystals. Not a single crystal of this mineral was found. This is odd in view of the rare but widespread occurrence of this mineral in marine K/T boundary rocks (Kyte and Smit, 1986).

One explanation, although unlikely in my opinion, for the lack of magnesioferrite crystals in continental K/T boundary rocks of western North America is that they were unstable and dissolved during diagenesis. Another more likely explanation is that the magnesioferrite crystals in marine K/T rocks are not of impact origin, but instead were derived from alteration of cosmic glassy spherules. Such spherules are known to accumulate in slowly deposited marine sediments (Brownlee, 1981, p. 743).

The thickness of the impact layer at most sites in the Raton basin ranges from 3.0 to 8.0 mm and averages 5.2 ± 1.1 mm. This average was obtained by studying the best 46 polished slabs of the K/T boundary interval from 12 sites. The thickest impact layer of this group, 8.0 mm, was measured on a polished slab from the Clear Creek North site in Colorado, although about one-half of this thickness consisted of authigenic jarosite laminations. At another site, not included in the above analysis of thickness, the impact layer is unusually thick. At the Carmel site west of Trinidad, Colorado, the impact layer varies from about 5 mm to as much as 20 mm. Where the impact layer is 20 mm thick, it contains abundant claystone clasts, graupen, and carbonaceous debris. This evidence suggests that the impact layer at this site is composed of a large amount of reworked material.

At other Western Interior North American K/T boundary sites, such as at the Teapot Dome and Sussex sites in Wyoming and at the Frenchman River site in Saskatchewan, the impact layer is locally thicker than it is in the Raton basin. For example, the thickness of the impact layer at the Teapot Dome and Sussex sites ranges from 3.0 to 10.0 mm, and at the Frenchman River site it is as much as 10 mm thick.

The texture of the impact layer claystone, as seen in thin sections, is different from that of the boundary claystone. It consists of alternating brown-stained laminae of kaolinite and feathery masses that probably are I/S mixed-layer clay because they have a markedly higher birefringence as compared to the kaolinite (Fig. 25). The orientation of the clay minerals in the feathery masses is roughly parallel to bedding. SEM analysis showed that, generally, the claystone has a texture characterized by a wavy open framework (Figs. 26B and 3B) typical of smectite and some mixed-layer clay. This texture is markedly different from the microspherulitic texture of the underlying boundary claystone (Fig. 4A).

Several lithic features make the impact layer a distinctive unit readily identifiable macroscopically and microscopically. The chief diagnostic feature of the claystone is its laminated nature, which contrasts markedly with the generally massive underlying boundary claystone (Fig. 11). Laminations of claystone are warped over and under clastic mineral grains, and this relation is indicative of soft-sediment compaction. The surfaces of the laminae display the impressions of macerated plant material, suggesting slow accumulation during normal depositional processes. Not only is the claystone laminated, but it also contains planar vitrinite laminae a few tens of micrometers thick. These vitrinite laminae probably formed from leaves and other plant materials.

Another diagnostic lithologic feature of the impact layer is its ubiquitous content of white to tan barley-shaped pellets of cryptocrystalline kaolinite similar to "graupen" of tonstein layers in coal beds (Williamson, 1970; Burger, 1985). In cross section they are eye- or almond-shaped (Fig. 26A), and in plan they are generally subcircular. Their major axes are parallel to bedding, and the aspect ratio is about 2, but some are more flattened and have a ratio of about 3. In general, planar laminae within the impact layer are deformed around many pellets, suggesting that the laminae were deformed by compaction around the more

Figure 25. Photomicrograph of a thin section of the K/T boundary impact layer. Specimen from the Berwind Canyon K/T boundary site, about 20 km northwest of Trinidad, Colorado, shows typical feathery texture of the claystone, laminae of black vitrinite, and gray almond-shaped kaolinite pellets. Scattered clastic grains of quartz and other minerals shown at arrows. Some of these grains contain multiple intersecting sets of shock lamellae. Photographed with cross-polarized light.

competent pellets. In some cases, a minute vitrinite shell encases the pellets. The claystone of many pellets appears massive and lacks any detectable structure such as a concentric or radial pattern. At the Carmel K/T boundary site west of Trinidad, Colorado, two different types of pellets or graupen occur in the impact layer. SEM study of the pellets showed that one type is composed of extremely fine-grained kaolinite, and the second type was formed of vermicular kaolinite similar to kaolinite pellets illustrated by Allen (1952, Fig. 1, Plate 2).

In addition to the ubiquitous barley-shaped graupen of cryptocrystalline kaolinite, the impact layer locally contains another type of small pellet (0.2 to 0.5 mm long) that has an irregular cross-sectional outline (Fig. 27). Some of these pellets have a microscopic texture identical to that of kaolinite of the boundary claystone. Some pellets contain wispy, macerated, carbonaceous plant material. Rare pellets contain minute clastic quartz grains. Such pellets are probably rip-up clasts of the underlying boundary claystone that were fairly well indurated prior to their erosion and redeposition.

At the Dogie Creek and Teapot Dome K/T boundary sites, the impact layer contains large (centimeter-size) kaolinite pellets that in cross section are generally lens shaped. Some of these pellets contain tiny goyazite spherules similar to those illustrated in Figure 6, irregularly shaped goyazite cavity fillings, and carbonaceous plant material. Bohor (1988a, 1988b) imagined that the pellets were originally impact-generated glass lapilli or bombs, but this cannot be the case because the pellets contain cellular carbonaceous plant material. In addition, the impact layer at the Dogie Creek site contains centimeter-long lenses of very fine-grained sandstone that are as much as 5 mm thick (Fig. 28A). Also at this site, the impact layer and subjacent boundary claystone are cut by sand-filled burrows (Fig. 27B) that possibly were made by insects (T. M. Bown, U.S. Geological Survey, oral communication, 1989).

Allen and Nichols (1945), Allen (1946, Figs. 2 and 5), and Grim and Allen (1938, Plate 1) described and illustrated kaolinite pellets in claystone beds and underclays similar to those in the impact layer. One occurrence is in the continental Calapooya Formation of Eocene age in Lane County, Oregon. In this formation, the pellets lack a radial or concentric structure, occur in a clay matrix, and are associated with charcoal, lignitic material, and diatoms. Allen and Nichols (1945) proposed that the pellets

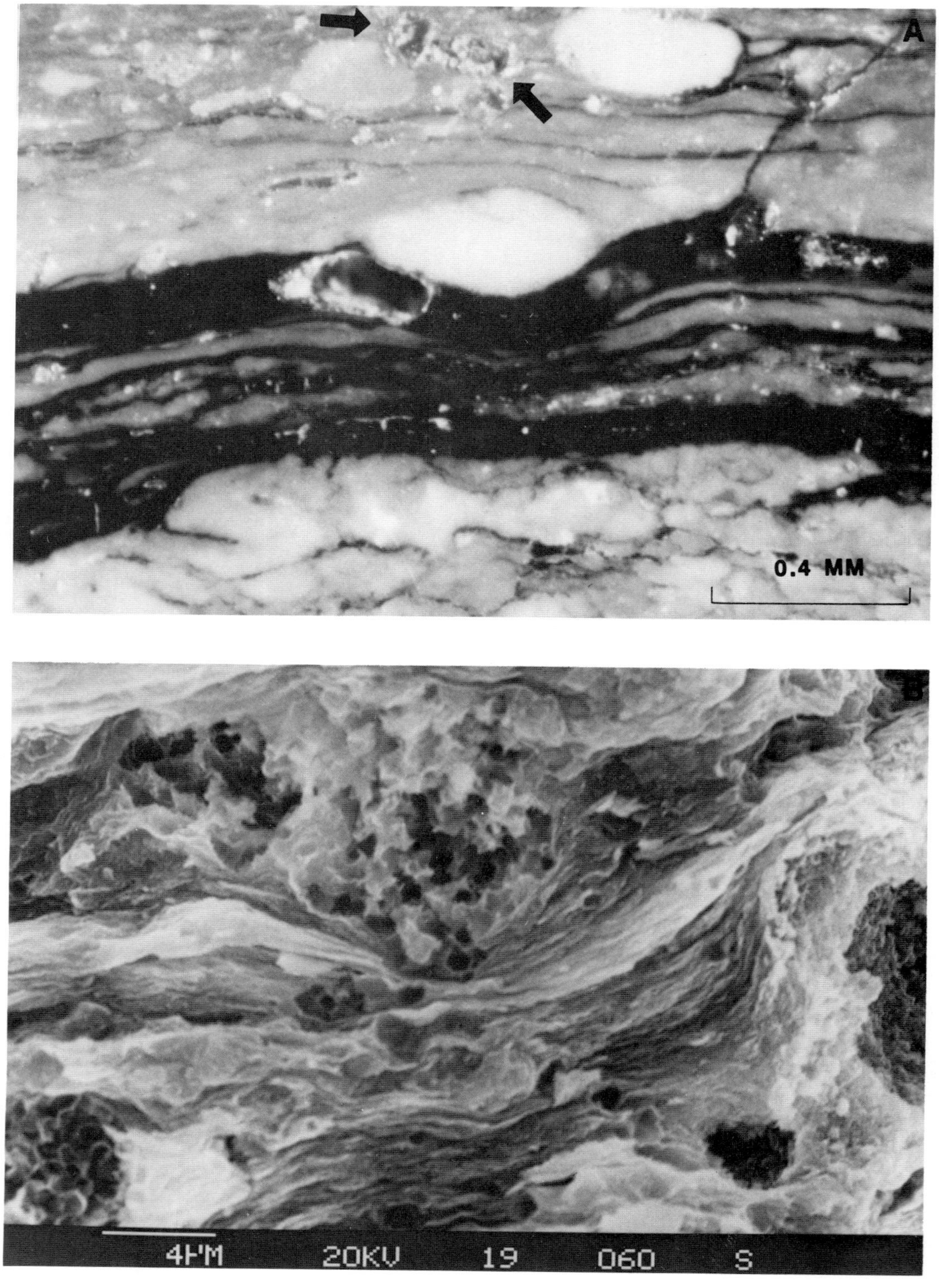

Figure 26. Photographs of the K/T boundary impact layer. A, polished surface of specimen from the Clear Creek North about 10 km south of Trinidad, Colorado, showing laminae of black vitrinite and white almond-shaped kaolinite pellets. Clastic quartz grains shown at arrows. Vitrinite laminae deformed below kaolinite pellet. B, SEM photomicrograph of specimen from the Madrid East K/T boundary site, about 10 km southwest of Trinidad, Colorado, locality showing the texture of the claystone. Note pits in claystone that were framboidal pyrite.

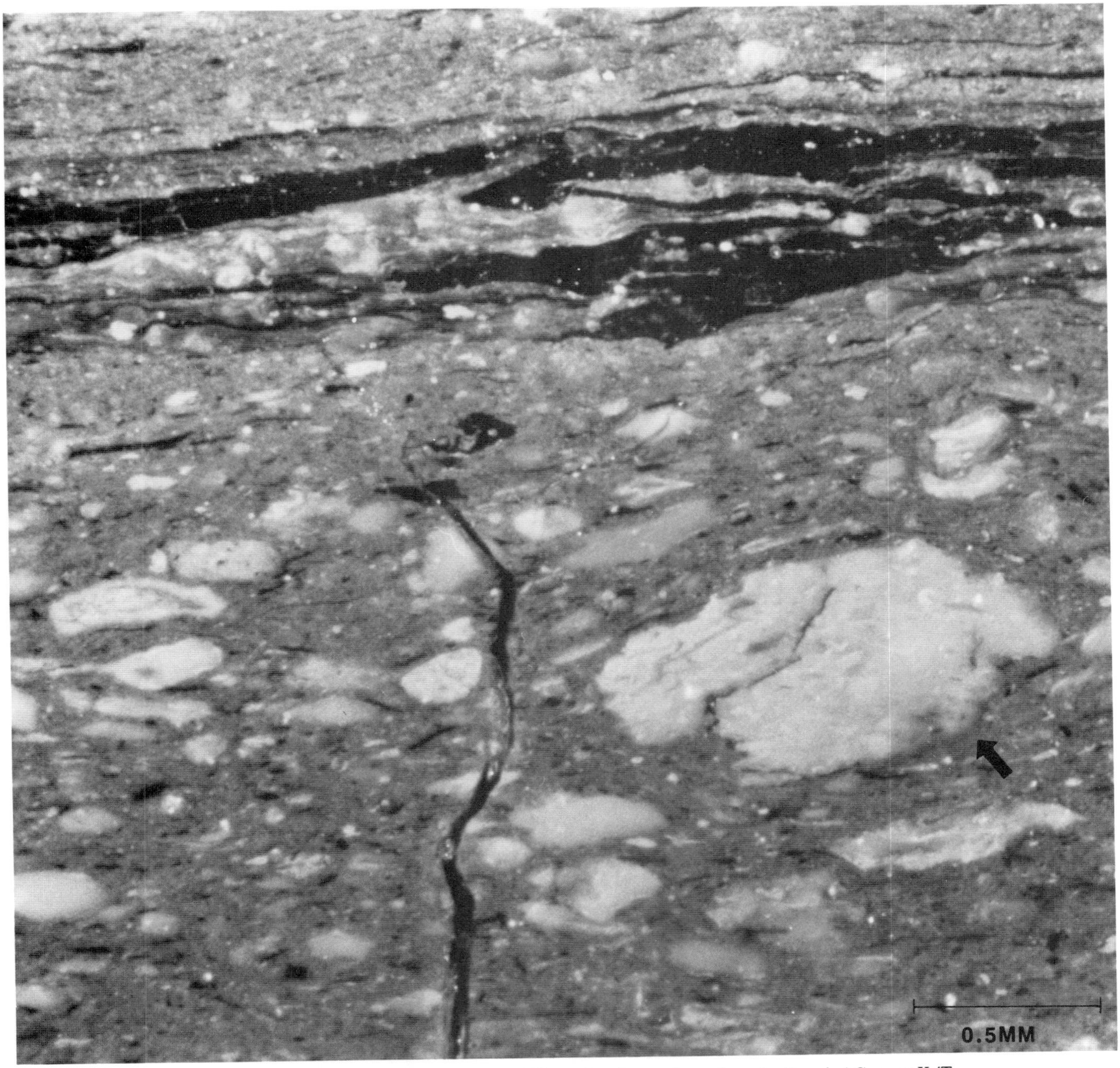

Figure 27. Polished surface of a slab of the K/T boundary impact layer from the Berwind Canyon K/T boundary site, about 20 km northwest of Trinidad, Colorado. Typical laminae of black vitrinite and light gray kaolinite pellets called "graupen" are shown. Large pellet at arrow has an irregular outline suggesting it is a rip-up clast of the underlying boundary claystone.

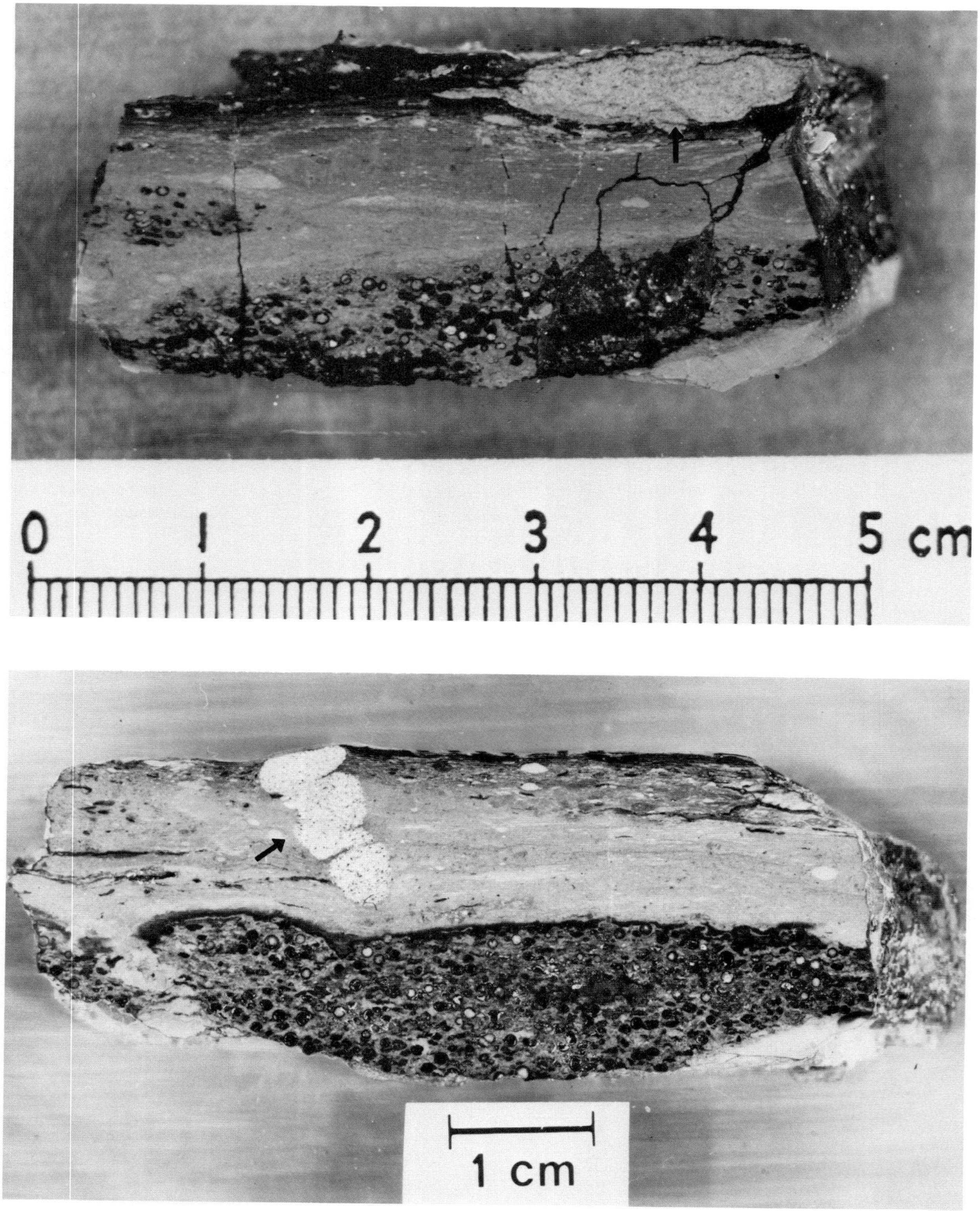

Figure 28. Photographs of polished slabs of the K/T boundary interval from the Dogie Creek, Wyoming, K/T boundary site. A, lens of very fine grained sandstone (shown at arrow) in the K/T boundary impact layer proves that it is composed, in part, of reworked material. B, very fine grained sandstone filling a burrow (shown at arrow) that cuts the boundary claystone and impact layer.

formed by a sedimentary-fluvial process whereby flakes of claystone were ripped up from the surface of Eocene flood plains and redeposited in the sediments that now form the Calapooya Formation. The pellets of the impact layer have a setting similar to those in the Eocene Calapooya Formation in that they occur in a clayey matrix and are associated with carbonaceous material. Because the pellets are similar to the kaolinite of the underlying boundary claystone, they may have formed during an episode of gentle, fluvial action that followed deposition of the boundary claystone. Kaolinite fragments that formed during desiccation and wetting of the uppermost surface of the boundary claystone could have been ripped up during a gentle fluvial episode and incorporated in the impact layer.

Another lithologic feature of the impact layer in the Raton basin is the widespread occurrence of minute jarosite spherules (10 to 50 μm) that probably formed from framboidal pyrite. In some samples, the jarosite spherules form planar groups and lenses near vitrinite laminae. As measured by the EDS analyzer of an SEM, some perfectly formed jarosite crystals of the spherules contain trace amounts of titanium and sodium. The titanium and sodium in the jarosite may account for its slightly different x-ray diffraction pattern as compared to that of typical jarosite. ICP analysis of the jarosite shows that it contains 0.3 percent titanium and 1,500, 120, and 360 ppm of arsenic, chromium, and vanadium, respectively (Table 5).

TABLE 7. X-RAY FLUORESCENCE ANALYSES OF THE K/T BOUNDARY IMPACT LAYER FROM COLORADO AND NEW MEXICO*

	BER	SVS	MADE	MR
SiO_2	42.2	46.8	45.7	42.7
Al_2O_3	26.0	25.3	28.9	27.5
Fe_2O_3	6.81	4.13	3.99	9.14
MgO	0.79	0.82	0.59	0.51
CaO	0.51	0.26	0.28	0.30
Na_2O	0.29	0.29	0.25	0.15
K_2O	2.42	2.47	1.38	1.24
TiO_2	1.42	1.29	2.27	1.81
P_2O_5	0.15	<0.05	0.06	0.17
MnO	<0.02	<0.02	<0.02	<0.02
LOI	18.0	18.3	16.7	16.9

*Analysts: A. J. Bartel, K. C. Stewart, and J. E. Taggart, U.S. Geological Survey; elements reported as oxides in weight percent; Fe calculated as Fe_2O_3; LOI, lost on ignition at 900°C. Localities of Figure 2: BER, Berwind Canyon (Colo.); SVS, Starkville South (Colo.); MADE, Madrid East (Colo.); MR, Madrid Railroad (Colo.); location of sites shown on Figure 2.

Chemical composition of the K/T boundary impact layer

The major-element chemical composition of four samples of the impact layer from the Raton basin was determined by XRF analysis, and the results are listed in Table 7. The analyses show that the impact layer is similar to the underlying boundary claystone (Table 2), except that the former contains considerably more iron and potassium than the latter, reflecting its jarosite (potassium-bearing hydrous iron sulfate) content. In addition, the claystone of the impact layer has a slightly larger Si to Al ratio than that of the boundary claystone.

The trace-element content of the impact-layer claystone, including jarosite laminae, was determined by the ICP method (Table 8), and its chemical composition is similar to that of the subjacent boundary claystone (Table 4) except that it contains less lithium and slightly more barium, chromium, copper, nickel, vanadium, and zinc. The trace-element content of the impact-layer claystone is also similar to a thin bed of black carbonaceous shale that overlies the K/T boundary impact layer at the Starkville South site in Colorado (Table 5). Moreover, the trace-element composition of the impact layer is similar, in general, to the average shale of the Earth's crust (Table 5).

The K/T boundary impact layer is similar to the underlying K/T boundary claystone in that they both have a low content of nickel (range 2 to 16 ppm) and cobalt (range 1 to 7 ppm), respectively (Tables 4 and 8). For comparison, a mudstone that occurs about 4 cm below the K/T boundary claystone layer contains an amount of nickel similar to that in the impact layer

TABLE 8. INDUCTION-COUPLED ATOMIC EMISSION SPECTROSCOPIC ANALYSES OF THE K/T BOUNDARY IMPACT LAYER FROM COLORADO AND NEW MEXICO*

Element	BER	BER	MR	MR	MADE
Ti	1.3	1.0	0.4	1.1	1.5
Mn	14	14	23	22	10
As	100	170	60	110	80
Ba	830	700	760	1200	330
Ce	78	58	180	52	20
Co	3	5	7	6	6
Cr	230	310	110	130	160
Cu	68	69	50	50	32
Ga	30	30	28	27	28
La	42	30	79	27	12
Li	89	51	15	97	81
Mo	8	6	10	4	8
Nd	31	23	82	26	9
Ni	6	6	16	9	6
Sc	40	43	24	24	30
V	280	320	190	170	60
Zn	45	52	73	2	37

*Analyst: M. J. Malcolm, U.S. Geological Survey; Ti in weight percent; other elements in parts per million. BER, Berwind Canyon (Colo.); MR, Madrid Railroad (Colo.); MADE, Madrid East (Colo.); SVS, Starkville South (Colo.); CCN, Clear Creek North (Colo.); location of sites shown on Figure 2.

(Tables 5 and 8). The low content of nickel and perhaps cobalt suggests that either the claystone of the impact layer does not contain a large component of asteroidal material or that these elements were removed during diagenesis. In contrast to the low content (<16 ppm) of these elements in the impact-layer claystone in continental K/T boundary rocks of the Raton basin, the K/T boundary impact layer in marine rocks at Caravaca, Spain, and Stevns Klint, Denmark, contains 2,580 (ppm) and 1,370 (ppm) nickel and 626 (ppm) and 146 (ppm) cobalt, respectively (Kyte and others, 1985).

In the Raton basin, Orth and others (1981) found and reported above-background amounts of iridium in continental sedimentary rocks at the precise palynological K/T boundary. Prior to this discovery, iridium abundance anomalies had only been found in marine rocks, and, thus, the discovery provided important support for the impact-extinction theory. Finding anomalously large amounts of iridium in continental rocks seemingly undercut suggestions by some that the K/T boundary iridium spike resulted from normal marine and geochemical processes, including (1) slow deposition and incorporation of iridium-bearing meteoritic dust or preferential removal of sediment at the expense of meteoritic material, thereby increasing the iridium content (Kent, 1981); and (2) precipitation of iridium from seawater owing to a change in ocean chemistry caused by the latest Cretaceous mass-extinction event (Kyte and others, 1981). If these suggestions were correct, iridium abundance anomalies should be found at numerous levels in the stratigraphic column. However, a methodical search for significant iridium abundance anomalies, other than the one at the K/T boundary, generally has been unsuccessful (Orth and others, 1985; Kyte, 1986; Orth and Attrep, 1988; Orth and others, 1989). However, a few small iridium anomalies have been found in marine rocks at the Frasnian and Famennian boundary in Australia and Morocco, in Lower Mississippian rocks of Oklahoma, at the Mississippian and Pennsylvanian boundary in Oklahoma and Texas, the Cenomanian and Turonian boundary in the Western Interior of North America, and in upper Eocene rocks (Orth and Attrep, 1988). With the exception of the upper Eocene iridium anomalies, these authors maintained that the iridium was concentrated by marine processes and is not of impact origin.

In their search for an iridium abundance anomaly near the K/T boundary in the Raton basin, Orth and others (1984) were able to concentrate their efforts in a narrow stratigraphic interval because of the palynological work of Tschudy (1973) and Pillmore (1976). Tschudy (1973) had identified a palynological break in the lower part of the Raton Formation, as well as at other places in the Western Interior of North America, and judged that the break marks the K/T boundary. As a result of the work of Orth and others (1984) and Pillmore and others (1984, Fig. 1), iridium abundance anomalies were found at eight sites in the sedimentary rocks of the Raton basin exactly at the palynological K/T boundary.

Because previous workers had not recognized or sampled the impact layer as a discrete unit, I made collections across the K/T boundary at four sites in the Raton basin to determine the iridium, silver, and gold elemental profiles. RNAA analyses of the samples were made by H. T. Millard, Jr., of the U.S. Geological Survey, following procedures he developed (Millard, 1987). The data are listed in Table 6 and shown graphically on Figure 15. Millard's analyses showed that the iridium content of the impact layer varied from 1.2 to 14.6 ppb; this amount was considerably more than that in the underlying boundary claystone and subjacent Cretaceous rocks. The detection limit for iridium, as determined by Millard, was about 0.004 to 0.07 ppb. In general, the iridium data are similar to those reported by Orth and others (1984), Pillmore and others (1984), and Gilmore and others (1984), but Pillmore and Flores (1987) reported a considerably larger iridium spike (56 ppb) at the Starkville South site in Colorado than that measured by Millard (1.2 ppb) at the same site. However, the data sets are not strictly comparable because the impact layer was not recognized or sampled by the above-cited authors.

As shown on Figure 15, the iridium concentration generally peaks in the impact layer; nevertheless, large values also can occur in carbonaceous-rich layers either below or above the impact layer. For example, significant iridium abundance anomalies have been measured in carbonaceous shale below the K/T boundary claystone at the Clear Creek North and Starkville South sites in Colorado (0.83 ppb and 1.5 ppb, respectively). At the Madrid East site near Madrid, Colordao, a significant iridium anomaly (0.47 ppb) was found in carbonaceous shale at the top of the boundary coal bed. Gilmore and others (1984, Fig. 2) also showed that a concentration peak of iridium occurred in coal or carbonaceous material slightly above the K/T boundary claystone bed at three of four sites they studied.

At the Carmel site near Madrid, Colorado (Fig. 2), a closely spaced set of samples that spans the K/T boundary was collected by the writer and Birger Schmitz of the University of Göteborg, Sweden, to determine their iridium content. Outcrops at this site in the Raton basin were chosen for additional detailed study because the K/T boundary claystone is locally not present (Fig. 10). The vertical line of section across the K/T boundary was chosen about 4.0 m from the west end of the outcrop at a place where the K/T boundary claystone was absent. Samples were collected at 1-cm intervals for 10 cm above and below the base of the K/T coal bed and at larger intervals above and below the +10 cm and −10 cm levels (Fig. 29). Samples were analyzed by B. Schmitz, H. Michel, and F. Asaro of the Lawrence Berkeley Laboratory of the University of California (B. Schmitz, written communication, 1989). The results of the analyses are graphically shown on Figure 29. The sample profile shows that iridium is concentrated in the K/T boundary impact layer (13.3 ppb), but significant amounts above background (about 3 to 70 parts per trillion) occur in coal or carbonaceous shale below and above the impact layer.

The surface concentration of iridium in the rocks that span the K/T boundary varies considerably at sites only a few kilometers apart (Figs. 15 and 29). Orth and others (1981) and Gilmore

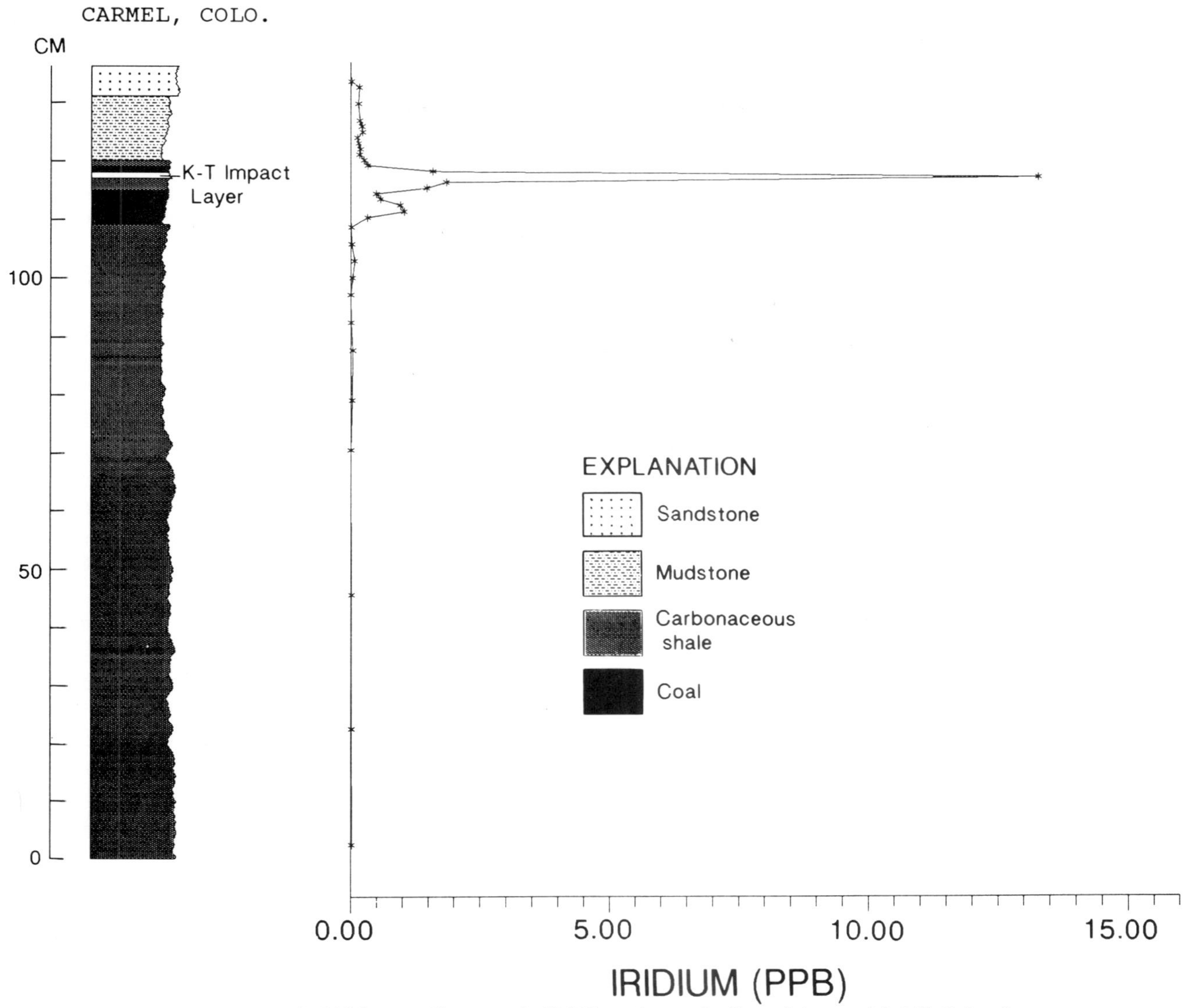

Figure 29. Detailed iridium profile across the K/T boundary at the Carmel site near Madrid, Colorado. Vertical line of sampling chosen at a place where the K/T boundary claystone is not present (see Fig. 10). Iridium is concentrated in the K/T boundary impact layer. Significant amounts of iridium also occur in subjacent and superjacent carbonaceous shale and coal. Iridium analyses by B. Schmitz, H. Michel, and F. Asaro of the Lawrence Berkeley Laboratory of the University of California.

and others (1984) reported amounts that average 50 ng/cm^2 and range from 8 to 120 ng/cm^2 at six sites in the Raton basin. Because shock-metamorphosed minerals are concentrated in the impact layer, I assumed that the impact-generated iridium also was originally concentrated in this layer. This assumption seems reasonable because Gostin and others (1989) showed that an iridium abundance anomaly occurs in a layer of Precambrian distal impact ejecta in South Australia. During diagenesis of the K/T boundary sediments, some iridium probably was mobilized and transported in solution away from the impact layer and concentrated in adjacent carbonaceous-rich sediments (Fig. 29). Bowles (1986) showed that platinum-group metals can be mobilized and transported in acidic low-temperature aqueous solutions and deposited in carbonaceous-rich sediment during weathering and diagenesis of rocks (see also Westland, 1981; Wallace and others, 1990).

Luck and Turekian (1983) determined the ratio of osmium isotopes ($^{187}Os/^{186}Os$) in a sample, presumably of the impact-layer claystone, collected at the Starkville South, Colorado, site by C. J. Orth, of the Los Alamos National Laboratory. Orth reportedly (in Luck and Turekian, 1983, p. 615) sampled material of the impact layer that contained the highest iridium abundance (Pillmore and Flores, 1987, Fig. 1) in the boundary interval, 56 ppb. According to Luck and Turekian (1983), the high-iridium claystone from Starkville South yielded a $^{187}Os/^{186}Os$ ratio of 1.29 ± 0.04. This ratio is much different than that of modern crustal material, which is about 10. They (1983, p. 615) observed that the ratio of osmium isotopes in claystone from the Starkville South site (1.29) is clearly different from the ratio (1.65) of osmium isotopes in marine boundary claystone (the so-called fish clay) from Stevns Klint, Denmark; so different that they represent asteroidal material of different compositions or different events. They also argued that the osmium isotopic ratio of the claystone is incompatible with the idea that the K/T asteroid was chondritic in composition. It also should be recognized that the impact layer in the Raton basin may not be the exact time equivalent of the material from the boundary claystone (fish clay) analyzed by Luck and Turekian (1983) at Stevns Klint, Denmark (see Fig. 23).

Origin of iridium anomaly

The origin of the anomalous amounts of iridium in K/T boundary rocks was attributed by Alvarez and others (1980, p. 1102) to "... an abnormal influx of extraterrestrial material." They suggested that a 10-km-size asteroid of chondritic composition struck Earth and created enormous amounts of dust-size crustal and asteroidal material. According to Melosh (1982, p. 124), most of the asteroid vaporized on contact with the Earth. The vaporized asteroid and its condensate, relatively rich in platinum-group elements, were ejected into the atmosphere and globally dispersed.

McLean (1985), Officer and Drake (1983, 1985), and Officer and others (1987) asserted that the source of the iridium at the K/T boundary was terrestrial and not extraterrestrial in origin. According to them, the iridium was generated during an intense period of volcanic activity in the Late Cretaceous and early Tertiary.

One possibility is that this volcanic activity was silicic. The evolutionary igneous process that begins with the generation of silicic magma and that ends with the ejection into the atmosphere of enormous amounts of pyroclastic material has been an often-repeated process throughout geologic time (Gilluly, 1965, p. 1). The record of silicic, pyroclastic volcanism in the Precambrian is known chiefly by the occurrence of clearly discernible metamorphosed ignimbrites and tuffs in Precambrian rock sequences. Precambrian pyroclastic rocks have been described in Arizona by Anderson and Creasey (1958); in Colorado by Olson (1976) and Wrucke and Dings (1979); in New Mexico by Barker and Friedman (1974); in the Midcontinent region by Hanson and Al-Shaieb (1980) and Sides and others (1981); and in Virginia by Rankin and others (1969, p. 741–744). It seems reasonable to assume that a considerable number of silicic ash beds must have resulted from these Precambrian pyroclastic volcanic eruptions.

Gilluly (1965, p. 1) noted that volcanism in the Phanerozoic was "... episodic but on the whole quasicontinuous." He further stated (1973, p. 504), "Except for two intervals in the Triassic, there have been no epochs as long as 5 m.y. without some record of siliceous magmatism in the North American Cordillera since the end of the Permian."

Numerous bentonites that record silicic, pyroclastic volcanism occur in Paleozoic marine rocks in the central and eastern United States. They have been reported in Ordovician rocks by Nelson (1921), Allen (1932), Bonnie and Honess (1929), Maddox (1930), Kay (1931), Rosenkrans (1936), Wilson (1949), and Ross and others (1981); in Silurian rocks by Honess (1923, p. 107–109); in Devonian rocks by Honess (1923, p. 123), Hass (1948), Fettke (1952), Flowers (1952), and Roen (1980); and in Mississippian rocks by Miser (1920), Honess (1921, p. 70–71), and Niem (1977).

An extraordinarily large number of bentonite beds occur in Mesozoic rocks of the Western Interior of North America. A biotite-bearing bentonite bed in marine Lower Triassic rocks (*Meekoceras* zone) was found by G. D. Frazer, of the U.S. Geological Survey (written communication, 1978), in the Pequop Mountains of eastern Nevada. Thin beds of bentonite have been reported from Upper Jurassic marine rocks in Colorado (Pipiringos and others, 1969, p. N-22), Wyoming (Pipiringos and Imlay, 1979, p. 23), and Utah (Schultz and Wright, 1963). Bentonite beds in the Jurassic Morrison Formation of the Colorado Plateau have been collected for K-Ar dating (J. D. Obradovich, oral communication, 1985).

It is well known that an exceptionally large number of bentonite beds occur in marine Cretaceous shale units of the Western Interior of North America (Rubey, 1929; Knechtel and Patterson, 1956; Slaughter and Earley, 1965; Hattin, 1975). For example, Kauffman (1985, p. 8) recorded about 700 discrete bentonite beds in marine rocks of the Western Interior of North America that range in age from Albian to late Maastrichtian. Pyroclastic materials in Cretaceous rocks in Arkansas and adjoining areas were reported by Ross and others (1928). Peaks of volcanic activity within the Cretaceous were reported by Heathman (1939, p. 16–17) who counted 71 bentonite beds in a roadcut exposure of 73 m of the Lower Cretaceous Mowry Shale near Casper, Wyoming. Mapel and Pillmore (1963) recorded 16 bentonite beds in the lower 23 m of the Mitten Black Shale Member of the Pierre Shale near Newcastle, Wyoming.

Thin tonsteins (altered volcanic ash) are common in coal-bearing continental Cretaceous and lower Tertiary rocks of Colorado, New Mexico, Utah, and Wyoming (Bohor, 1977; Ryer and others, 1980), but they are not nearly as common as bentonite beds in marine Cretaceous rocks. They can be found in Paleocene and Eocene coal beds in Colorado, New Mexico, Utah, and Wyoming (Bohor, 1977). Altered volcanic ash beds in the Eocene Green River Formation in Colorado, Utah, and Wyoming were studied by Bradley (1964) and Desborough and others (1973). Altered volcanic ash beds are common in Eocene and Oligocene sedimentary rocks in the Gulf Coast states, according to Ross and others (1928, p. 193).

Lenticular layers of glassy volcanic ash are ubiquitous in upper Tertiary sedimentary rocks of the Western Interior of North America. Volcanic ash beds are interbedded in sedimentary rocks of the Great Plains including the White River (Oligocene), Arikaree (Oligocene and Miocene), and Ogallala (Miocene) Formations and their equivalents in the Rocky

Mountains and Great Basin (Izett and Barclay, 1973; Galusha and Blick, 1971). These ash beds attest to repeated episodes of pyroclastic volcanism in the Great Basin (Evernden and others, 1964; Izett, 1968). About 70 silicic ash beds were reported by Izett (1981) in Pliocene and Pleistocene sedimentary rocks of the Western Interior of North America.

From the foregoing discussion it is clear that silicic pyroclastic volcanism has been an often-repeated process throughout Precambrian and Phanerozoic time, although the geologic record does not support the idea that there was an intense period of silicic volcanism at the K/T boundary. Altered silicic ash beds (bentonites) are abundant in Cretaceous marine rocks of western North America and represent several peaks of volcanic activity (Izett, 1981, p. 10202–10203), but they are rare in continental K/T boundary interval rocks of North America. Not only are silicic volcanic ash beds rare in the K/T boundary interval rocks of North America, but also in marine K/T boundary interval rocks elsewhere in the world, including Europe, New Zealand, Alaska, and Asia. The paucity of radiometric ages from rocks near the K/T boundary reflects the general global absence of datable silicic volcanic rocks in this interval. Of more importance is the fact that silicic volcanic rocks are known to have extremely low (<0.030 ppb) iridium concentrations (Orth and Attrep, 1988; Orth and others, 1989), and thus they cannot be the source of the iridium anomaly at the K/T boundary.

A second possibility is that the intense period of K/T boundary volcanism proposed by McLean (1985), Officer and Drake (1983, 1985), and Officer and others (1987) was basaltic; indeed, basaltic rocks contain more iridium than silicic ones (Palme, 1982; Orth and Attrep, 1988). Moreover, relatively large amounts of iridium were detected in airborne particulate material ejected from Kilauea, a classic basaltic volcano on the island of Hawaii (Zoller and others, 1983; Olmez and others, 1986). Furthermore, it is true that an enormous volume of basalt, the Deccan Traps, was extruded in India during the Late Cretaceous and early Cenozoic, but there has been considerable debate about the precise radiometric age for the sequence of flows. Baksi (1987) contended that the bulk of the Deccan Traps were emplaced during the interval from 70 to 60 Ma. On the basis of paleomagnetic data, Stoddard and Jurdy (1988) estimated that the Deccan Traps were erupted from 83 to 70 Ma. Courtillot and others (1986, p. 362) stated that the ages of the Deccan basalt flows range from 85 to 30 Ma. Thus, the volcanic activity probably would not produce a sharp iridium anomaly at exactly 66 Ma. Duncan and Pyle (1988) made new $^{40}Ar/^{39}Ar$ ages of a thick section of Deccan basalts from western Ghats. They concluded that the basalts range in age from 66.6 to 68.5 Ma, slightly older than the K/T boundary. Courtillot and others (1988) concluded that, on the basis of the most reliable $^{40}Ar/^{39}Ar$ radiometric ages, the bulk of the Deccan basalts were erupted during a 4-m.y. period from 65 to 69 Ma. Besse and others (1989) reevaluated all previous geochronologic and paleomagnetic information relative to the Deccan Traps and decided that their age is in the range 70 to 65 Ma. Baksi (1989) also reevaluated previous geochronologic work on the Deccan Traps and concluded that they were extruded during a 2- to 3-m.y. period near 65 Ma. Thus, there does not seem to be a spike of basaltic volcanism at ~66 Ma coeval with the iridium spike in K/T boundary continental and marine rocks.

Setting the age problem aside, there are objections to the idea iridium at the K/T boundary was of basaltic volcanic origin. As is well known, basaltic volcanism is not explosive, and the products of such volcanic activity are not transported far downwind from their eruption sites. Izett (1981) studied regionally distributed upper Cenozoic volcanic ash beds of the Western Interior of North America, and nearly all are of silicic composition. For example, no basaltic ash beds have been found in upper Cenozoic rocks of the Western Interior of North America that are far downwind from the middle Miocene Columbia River basalt field. In brief, particulate airborne products of basaltic volcanism are found only near their vents.

A much more serious objection to the idea that the K/T iridium anomaly is volcanic in origin is that shocked quartz grains have been found globally at the K/T boundary closely associated with the iridium anomaly. These shocked-quartz grains cannot be a product of basaltic volcanism, and as will be argued elsewhere in this paper, the particular types of shocked quartz in the K/T boundary impact layer cannot be of volcanic origin. Even if the shocked quartz were of volcanic origin, there is no known common geologic process that could atmospherically distribute large grains (~100 μm) globally. In spite of this generalization, Betzer and others (1988) recently found quartz grains as large as 200 μm in dust traps in the north Pacific ocean 10,000 km downwind from their source; presumably, they were transported atmospherically following a major dust storm in China.

On the basis of the discussion in the foregoing paragraphs, the hypothesis of Alvarez and others (1980) that the iridium anomaly at the K/T boundary is of extraterrestrial origin is reasonable and in accord with Occam's razor in light of the established impact record of the Earth (Shoemaker, 1979; Shoemaker and others, 1988) and the fact that iridium anomalies can occur in distal impact ejecta. Gostin and others (1989) detected an iridium anomaly in distal impact eject derived from the Precambrian Acraman impact structure in South Australia.

Origin of the K/T boundary impact layer

Evaluation of stratigraphic data bearing on the origin of the K/T boundary impact layer might lead to the conclusion that the claystone is formed entirely of primary air-fall impact-derived material. The main stratigraphic observation that could be used to support this conclusion is that the impact layer forms a discrete thin unit, 5 to 10 mm thick, that can be identified at about 30 scattered sites in western North America. At least 20 sites have been found in the Raton basin alone. In addition to the Raton basin sites, the impact layer occurs in Wyoming at the Teapot Dome, Dogie Creek, and Sussex sites. Much farther north, the impact layer was identified in polished slabs of K/T boundary interval rocks, including sites at (1) Brownie Butte, Montana; (2) Morgan Creek, Saskatchewan, Canada, of Nichols and others (1986); (3) Frenchman River, Saskatchewan, Canada (Lerbekmo and others, 1987); and (4) Red Deer Valley north of Calgary, Canada (Lerbekmo and St. Louis, 1985). Thus, the widespread and thin nature of the layer seems to suggest that it is composed of air-fall material.

In contrast, observational information indicates that the impact layer has a complex origin and is composed of no more than 2.0 percent by volume of impact materials, principally shocked and unshocked target materials. My observation that the impact layer is composed of finely laminated claystone that contains vitrinite laminae implies that it did not form by simple diagenetic alteration of coarse-grained impact glass, but rather by slow accumulation of clay particles. The texture of much of the impact-layer claystone, as determined by SEM analysis, is similar to mixed-layer clay and is not pervasively microspherulitic like the boundary claystone. This fact also suggests that the claystone was not derived from volcanic or impact glass, if the contention of Pollastro and Pillmore (1987) were correct that the alteration of glass to kaolinite results in a microspherulitic texture. The occurrence of macerated plant detritus and twigs along bedding laminae implies, but does not prove, that the claystone accumulated slowly by fluvial or paludal, not air-fall, depositional processes. The presence in the impact layer of sandstone lenses, kaolinite pellets, and graupen of clastic origin suggests that the impact layer is mainly constituted of water-reworked material.

Nonclay mineral concentrates of the impact layer, described in a subsequent section of this paper, contain a miniscule amount of small (<100 μm) rounded zircon, tourmaline, epidote, and other heavy minerals that clearly are not of impact origin. Similar stable heavy minerals occur in sedimentary rocks stratigraphically below and above the K/T boundary. The presence of these heavy minerals in the impact layer strongly implies that the unit is either not composed of primary air-fall impact material or has been pervasively contaminated with locally derived sediment.

The trace-element content (platinum-group metals excluded) of the impact layer indicates that it does not contain a large component of asteroidal material. The low content of nickel, in particular, is relevant to this conclusion.

In summary, observational evidence seems to suggest that the claystone of the impact layer in the Raton basin formed principally from locally derived sediment and was not formed from altered air-fall volcanic or impact glass. Others have concluded that the impact layer in marine K/T boundary rocks is likewise composed of normal marine sediments and is not composed of altered impact target material (Rampino, 1982; Rampino and Reynolds, 1983; Johnsson and Reynolds, 1986; Schmitz, 1988; Elliott and others, 1989). Apparently, the impact-layer claystone contains only a minor fraction of impact-derived material, principally shocked and unshocked minerals. If this conclusion is correct, the thickness of the impact layer does not vary as a function of distance from the K/T impact site.

If the claystone of the K/T impact layer in continental rocks of western North American sites is not composed of altered glassy air-fall impact material, the natural question that arises is whether the same interval of rock at marine sites in Europe is composed of altered glassy air-fall material. Many geologists believe that the K/T boundary claystone layers at the classic sites of Stevns Klint, Denmark, and Caravaca, Spain, are composed of altered air-fall glassy material. This myth is not supported by direct observation of the rocks at these sites.

A photograph of a polished slab of the K/T boundary interval from the Stevns Klint, Denmark, site is shown on the back cover of this paper. The rock unit above the white Cretaceous marl (units 2 and 3 of Christensen and others, 1973) is part of the so-called "fish-clay," and it contains the principal iridium anomaly and rare shock-metamorphosed quartz grains (Kyte and others, 1980; Smit and Kate, 1982; Bohor and Izett, 1986; Schmitz, 1988). The unit is composed of microlaminated brownish gray marl that contains abundant flakes and shreds of carbonaceous material oriented parallel to bedding. The marl also contains small (1.0 mm), angular clasts of glauconite and altered and partially altered pyrite spherules. The brownish gray marl is interlaminated with light gray lenticular laminae that contain slighlty more calcium carbonate than the brownish gray marl. Thus, the boundary interval at Stevns Klint is composed of reworked material that obviously does not resemble discrete layers of air-fall material (bentonites) in marine rocks. More than a decade ago, Christensen and others (1973, p. 209) pointed out that the marl of the so-called fish clay is of ". . . terrigenous origin, washed out into the sea from nearby land areas." This information largely has been ignored by those who champion the air-fall impact or volcanic origin for the boundary claystone.

A depositional model for the origin of the impact layer in the Raton basin that fits most of the observational evidence gathered in this study is as follows. Vaporized and melted asteroidal material and shocked and unshocked target materials were produced during the K/T boundary impact event. A small amount of fine-grained shocked mineral grains and an iridium-rich aerosol were transported away from the impact point by upper stratosphere winds. Only a sprinkling of shock-metamorphosed minerals, some of which were relatively large (0.5 to 0.7 mm), and the iridium-rich aerosol fell throughout much of North America, but the materials were only preserved in exceptionally low-energy environments in coal basins that existed in the Raton basin of Colorado and New Mexico and elsewhere in the Western Interior of North America. The materials fell on a paleosurface that existed at the top of the K/T boundary claystone. Locally, plants grew on this surface, and the surface may have been covered intermittently with water. J. A. Wolfe, of the U.S. Geological Survey has identified abundant fossil lily pads at the Teapot Dome, Wyoming, K/T boundary site (written communication, 1989). Although the impact layer may have formed in shallow water, no sedimentary structures, such as ripple marks, have been found in the unit.

The hydrologic regime in the Raton basin may have been

slightly upset by climate change as a result of the impact event, which triggered a new but amazingly gentle erosional-depositional cycle. During this cycle, shock-metamorphosed minerals were reworked gently into a slowly accumulating deposit composed of clay and decaying plant material. Locally, rip-up clasts of kaolinite formed in the impact layer. Because there is only one spike of shocked minerals found to date in the K/T boundary interval, there is no direct physical evidence to support the idea (Hut and others, 1987) that there were multiple impacts during this time. Finally, the data also suggest that the K/T boundary impact event was considerably smaller than had been proposed by Alvarez and others (1980).

ROCKS DIRECTLY OVERLYING THE K/T BOUNDARY IMPACT LAYER

A coal bed, informally called the "K/T boundary coal" (Fig. 12), directly overlies the impact layer at most K/T boundary sites in the Raton basin and at other sites in western North America (Izett, 1987c, 1988). The association of a coal bed with the K/T boundary claystone provides an important tool for prospecting for new K/T boundary sections. Obviously, this coal bed does not extend continuously from the Raton basin to southern Alberta and Saskatchewan, Canada. Nevertheless, coal-forming conditions existed at the exact stratigraphic horizon of the K/T coal bed over an amazingly large area from New Mexico to southern Canada.

The coal beds in the Vermejo and Raton Formations of the Raton basin are mostly low-sulfur bituminous A to B in rank (Pillmore and others, 1984). In contrast, age-equivalent coal beds in Wyoming, Montana, and southern Canada are of lower rank, mainly lignite. At fresh exposures the coal has a bright luster and a well-defined cleat. Fracture surfaces in the coal are commonly coated with jarosite. Thin sections of the coal show that it contains extremely rare clastic mineral grains, chiefly quartz.

The coal ranges in thickness from 1 to 16 cm; it is thickest (16 cm) at the Sugarite site in New Mexico and thinnest (1 cm) at the Carmel site in Colorado. To the north in Wyoming, Montana, and Canada, the coal bed ranges in thickness from about 4 cm at the Dogie Creek, Wyoming, site (Bohor and others, 1987) to as much as 75 cm at the Hell Creek, Montana, site (Baadsgaard and others, 1988, p. 1089).

The contact between the impact layer and the overlying boundary coal bed is sharp and occurs within a 1-mm-thick interval that separates laminated claystone from coal (Fig. 1). At some sites in western North America, the coal or lignite just above the contact is shaly and contains sparse kaolinite graupen.

At two sites in the Raton basin of New Mexico—Climax Canyon and Raton—an anomalous stratigraphic situation occurs. At these sites, the impact layer and K/T boundary coal bed are separated by an intervening claystone 1 to 23 cm thick. At some places in the western excavation pit at the Raton site, the impact layer and boundary coal bed are separated by an intervening greenish gray claystone as much as 6.0 cm thick. Locally, the base of this unit contains lenses of very sandy mudstone. At other places in the excavations, the impact layer is directly overlain by a thin coal seam about 0.2 cm thick, which in turn is overlain by 1.0 cm of claystone and the boundary coal bed. These two sites are the only places in the Raton basin where the boundary claystone and impact layer are not directly overlain by coal. Similarly, a unit of intervening claystone that ranges from 4 to 10 cm thick locally occurs between the impact layer and the K/T boundary coal (lignite) bed at the Teapot Dome, Dogie Creek, and Sussex, Wyoming, sites.

The trace-element chemical content of the coal was determined by the ICP method, and the results are listed in Table 5. Inspection of the data shows that the coal is impoverished in the analyzed elements as compared to other boundary interval rocks, except for zinc.

The iridium content of the boundary coal can nearly equal that of the underlying impact layer at some boundary sites and can exceed that in the impact layer at other sites. For example, the iridium content of the impact layer is only 1.2 ppb at the Raton site. In contrast, the boundary coal contains about 10.9 ppb iridium at the Madrid site in Colorado (Fig. 15). One explanation for the iridium anomaly in the coal is as follows. During diagenesis of impact-layer sediments, iridium was chemically mobilized and moved by ground water into the adjacent coal where it was fixed (see Wallace and others, 1990). An argument could be made that impact materials were moved upward from the impact layer into the coal by mechanical sedimentary processes, but there is no evidence for such a reworking event. In particular, the coal only contains exceedingly rare quartz and other clastic mineral grains.

CLASTIC MINERAL GRAINS OF THE K/T BOUNDARY IMPACT LAYER

Clay-free mineral-grain concentrates of K/T boundary interval rocks from the Raton basin and other western North America K/T boundary sites were prepared to determine their mineralogy. I was interested in the composition of the various mineral phases, particularly shock-metamorphosed minerals, that compose the concentrates. The size and abundance of the shocked mineral grains in rocks at these sites relative to the same properties in rocks at sites outside North America was of great interest to me. I reasoned that if the composition, size, and abundance estimates could be determined, the data might be used to narrow the search for the location of the K/T boundary impact.

The method used was as follows. The exact stratigraphic location of the shock-metamorphosed materials in the K/T boundary interval in the Raton basin was established by studying polished slabs of boundary-interval rocks mounted in plastic and cut normal to bedding. A concentration of relatively large mineral grains including quartz was found to occur in the K/T boundary impact layer (Izett and Pillmore, 1985a, 1985b; Izett, 1987c, 1988). Grains were plucked from the polished surfaces of the impact-layer claystone, and some of the grains were found to contain multiple intersecting sets of shock lamellae. Subsequently, shock-metamorphosed grains were identified in thin sections of the impact layer, and numerous gram-size samples of shock-metamorphosed material were obtained for study by dissolving the clay minerals in kilogram-size samples.

Preparation of clay-free mineral concentrates from impact-layer samples was accomplished by (1) collecting pieces of the boundary interval composed of both the boundary claystone and the impact layer, (2) scraping off the impact layer from the top of the boundary claystone with a razor blade, (3) pulverizing the impact-layer material in water in a cocktail blender, (4) dissolving this slurry in a hydrofluoric acid solution (~5 percent) accompanied by ultrasonic cleaning or scrubbing, and (5) decanting the suspended material, mostly clay minerals. The residuum again was ultrasonically cleaned, but this time in an aqueous solution of household bleach (~30 percent). The residuum was reduced further in volume by dissolution of the remaining clay minerals in dilute (~5 percent) hydrofluoric acid. Grains of authigenic gypsum, barite, rutile, and jarosite survived the original acid digestion, and of these, gypsum and jarosite were removed by further acid digestion (HCl and H_2SO_4). Barite and rutile were removed using heavy liquid separation methods.

The assemblage of mineral grains separated from K/T impact-layer claystone samples first was studied with a biologic-type stereomicroscope and second with a petrographic microscope. Grains suspected to be of shock-metamorphic origin because of their appearance under 90× magnification of a biologic-type stereomicroscope were mounted on the tip of a steel-wire spindle. The mounted grains were inserted into the immersion cell of a spindle stage (Wilcox, 1959). This device is ideal for the study of shock-metamorphic features in mineral grains because planar shock lamellae and other planar optical features can be rotated vertical, and their orientation in relation to optic directions can be measured and plotted on a stereonet. The spindle stage is also ideally suited for the measurement of the principal indices of refraction of shocked minerals. Using the spindle stage, the precision of measuring planar features and optical directions is about ± 2° (Wilcox, 1959).

Shock-metamorphosed minerals

Microscopic study of mineral concentrates obtained from the impact layer at sites in the Raton basin and elsewhere in western North America showed that they consisted of (1) single grains of quartz, chert, chalcedony, oligoclase, and potassium feldspar; and (2) composite grains (lithic fragments) composed of metaquartzite, quartzite, and granite-like aggregates of quartz, oligoclase, and potassium feldspar (microcline). Some of these contain shock-metamorphic features.

Quartz grains are the chief component of the clastic mineral grains in the impact-layer claystone. Many of the grains appear structureless and show normal sharp extinction when examined with a petrographic microscope using cross-polarized light. Some grains show anomalous extinction, and others exhibit Böhm lamellae. Quartz constitutes from 30 to 82 percent (mean = 50 percent) of the assemblage of clastic mineral grains in the K/T boundary impact layer at 12 Western Interior North American K/T bundary sites (Table 9). The average content of quartz in the total grain assemblage in eight samples from the Raton basin is 43 percent. Significantly, about 50 percent of these, or 21 percent of the total assemblage, are shock-metamorphosed; that is, they contain multiple sets of closely spaced planar shock lamellae.

At two sites, one in Colorado and a second in Wyoming, the assemblage of clastic grains contains an uncommonly large amount of feldspar and quartz. At the Madrid East site, near Madrid, Colorado, the assemblage of grains contains an unusually high concentration of feldspar (17 percent), mostly microcline (Table 9). The high content of microcline and the low content of shocked quartz at this site compared to other sites suggest that the microcline is of local derivation and that the impact layer consists, in part, of reworked material. At the Dogie Creek, Wyoming, K/T boundary site, the impact layer contains more than 80 percent quartz grains, and only 17 percent of these are shocked (Table 9). The unusually high content of unshocked quartz at this site, as compared to other sites, suggests that the impact layer consists of reworked material, as discussed in a previous section of this paper.

As viewed under a biologic-type stereomicroscope, some quartz grains appear cloudy or milky and have a pearly luster; these features provide a clue to their shock-metamorphic origin. At about 90× magnification, a few quartz grains have sets of planar lamellae that also provide a clue to their shock-metamorphic origin. The planar lamellae of such grains are similar in some respects to polysynthetic twin lamellae in plagioclase. In some instances, the hydrofluoric acid selectively etches the shock lamellae, making them especially obvious.

As seen with a petrographic microscope using plane-polarized light, quartz grains with multiple sets of planar lamellae are either colorless or tinted very pale brown. Figure 30 shows five quartz grains separated from the impact layer in the Raton basin and one from Brownie Butte, Montana, that show multiple sets of planar lamellae. For example, the grain illustrated in Figure 30A has four sets of shock lamellae in this grain orientation. The shock lamellae are typically planar and unlike lamellae formed by tectonic processes. It cannot be overemphasized that quartz grains such as those illustrated in Figure 30 are typical and common in the K/T boundary impact layer residues, and they provide a large amount of shocked quartz for study.

Quartzite and metaquartzite grains form a significant proportion of the clastic grains in the impact layer at Western Interior North American K/T boundary sites. The distinction between quartzite and metaquartzite is somewhat arbitrary. Grains classified as quartzite are distinctly granular, have fairly distinct grain borders, and are fine grained (0.05 to 0.1 mm). Some grains classified as metaquartzite are very fine grained (0.01 mm) and have optical boundaries that vary from curved to highly sutured. Rare grains are slightly micaceous, foliated, and similar to sericitic quartz schist.

TABLE 9. COMPOSITION AND ABUNDANCE OF MINERAL GRAINS, SOME OF WHICH CONTAIN MULTIPLE SETS OF SHOCK LAMELLAE, IN THE K/T BOUNDARY IMPACT LAYER AT WESTERN NORTH AMERICAN K/T BOUNDARY SITES*

	CCN	MADE	MR	BER	CAR1	CAR2	LC	RAT	TD	DC	BB	FR
Quartz (Unshocked)	12	39	38	12	16	27	20	10	27	65	13	36
Quartz (Shocked)	20	8	30	24	31	10	28	20	42	17	28	25
Quartzite-Metaquartzite (Unshocked)	27	24	14	17	20	21	24	20	15	7	24	15
Quartzite-Metaquartzite (Shocked)	9	2	6	15	8	6	6	13	5	1	5	4
Feldspar (Unshocked)	1	17	0	0	0	0	0	1	0	2	13	2
Feldspar (Shocked)	1	2	0	0	0	0	0	0	0	0	3	1
Chert	28	8	9	29	25	36	20	33	9	5	14	16
Chalcedony	2	0	3	3	8	10	2	3	2	3	0	1

*100 randomly selected grains were mounted individually on the tip of steel needle and studied microscopically with a spindle stage. Amount of constituent in percent. CCN, Clear Creek North (Colo.); MADE, Madrid East (Colo.); MR, Madrid Railroad (Colo.); BER, Berwind Canyon (Colo.); CAR1, Carmel (Colo.) at a place where K/T boundary claystone present; CAR2, Carmel (Colo.) at a place where K/T boundary claystone *not* present; LC, Long Canyon (Colo.); RAT, Raton (N. Mex.); TD, Teapot Dome (Wyo.); DC, Dogie Creek (Wyo.); BB, Brownie Butte (Mont.); FR, Frenchman River, Saskatchewan (Canada); location of sites shown on Figures 2 and 45.

A few percent of these grains or lithic fragments exhibit shock-metamorphic features. Individual quartz optical units that compose a lithic fragment contain multiple intersecting sets of shock lamellae. Typical grains of shock-metamorphosed quartzite and metaquartzite from the impact layer are shown in Figures 31 and 32. Shock lamellae in each optical unit parallel crystallographic planes of the quartz lattice and terminate at optical grain boundaries. These observations are in accord with data from quartz at meteor impact sites (Carter and Friedman, 1965). Such grains also have been seen in K/T boundary mineral concentrates from other Western Interior North American sites and at the Caravaca, Spain, and Stevns Klint, Denmark, sites in Europe (Fig. 33).

Quartzite and metaquartzite make up from 8 to 36 percent of the grains in the impact layer at Western Interior North American K/T boundary sites (Table 9). In the Raton basin they range from 20 to 36 percent (mean = 29 percent). In samples from the Raton basin, from 8 to 47 percent (mean = 27.5 percent) of the quartzite and metaquartzite grains (~8 percent of the total population) appeared to contain shock lamellae during spindle stage study (Table 9), but this is only a crude estimate because shock lamellae are difficult to see in these polycrystalline grains. In thin sections, sets of shock lamellae in the quartzite and metaquartzite grains are more easily seen and could be counted, but not all sets of shock lamellae can be recorded because of the inability to rotate grains during point-counter analysis.

Most of the quartzite and metaquartzite grains do not appear to contain shock lamellae, and thus some of the grains could be normal unshocked clastic mineral grains incorporated in the impact layer during mixing and reworking with local sediment. However, the fact that some of the grains contain shock lamellae indicates that the source for some of the grains was atmospherically transported impact-target rocks. Moreover, the greater abundance of quartzite and metaquartzite grains in the impact layer at most K/T sites relative to their abundance in underlying rocks (see Tables 1 and 9) suggests that the provenance of most of the unshocked grains was not from the normal erosion of older rocks but rather from impact-target rocks.

Chert and chalcedony compose from 8 to 46 percent (mean = 27 percent) of the grains in the impact layer in the Raton basin (Table 9). Microscopic study of thin sections of these minerals shows that, in general, they have different textures, but there is some gradation between these two materials. Under a petrographic microscope using cross-polarized light, most chert grains consist of a texturally uniform aggregate of interlocking granules no more than a micrometer in diameter. Some chert grains contain patches of coarser grained material. Chalcedony grains, on the other hand, are coarser grained than chert, and they consist of an irregular patchy mosaic of interlocking grains. However, some grains consist of several domains of different grain size. The patchy extinction under cross-polarized light of some chalcedony grains is similar to the patchy extinction of some shock-metamorphosed quartz grains. Some chalcedony grains are spherical, and when viewed under a petrographic microscope using cross-polarized light, they display a dark, pseudouniaxial interference figure that rotates as the microscope stage is turned. Typically, some chalcedony grains contain scattered opaque cubic crystals, as large as 20 μm, that the EDS analyzer of an

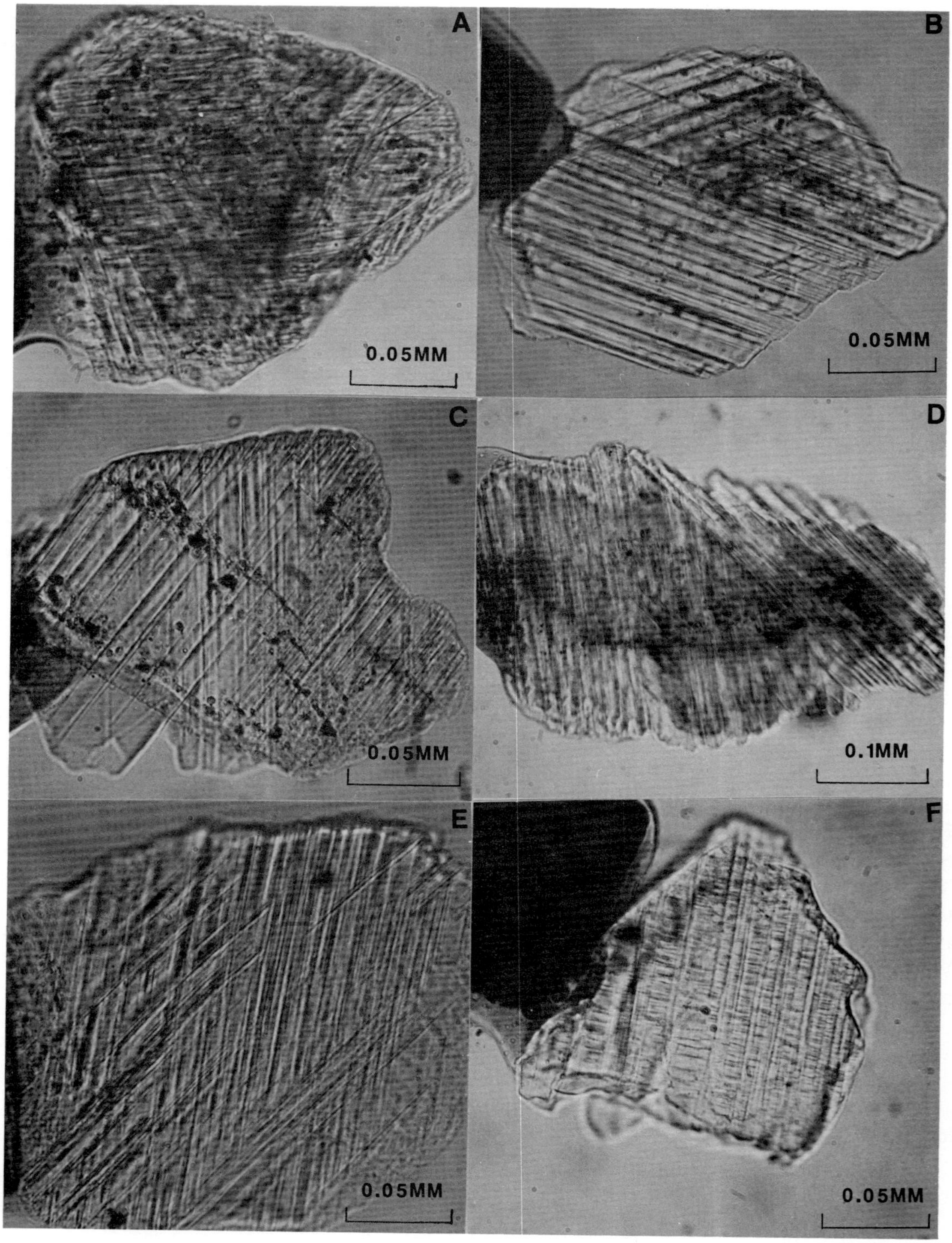

Figure 30. Photomicrographs of shocked quartz grains from the K/T boundary impact layer. A, grain from the Clear Creek North site, showing four sets of shock lamellae; B, grain from the Brownie Butte, Montana, site, showing two sets of shock lamellae; C, grain from the Starkville South site locality, showing two sets of shock lamellae and numerous opaque inclusions; D, grain 0.42 mm long from the Clear Creek North site, showing two sets of shock lamellae; E, grain from the Starkville South site, showing two sets of shock lamellae; F, Grain from the Clear Creek North site, showing two closely spaced sets of shock lamellae. Photographed with plane-polarized light.

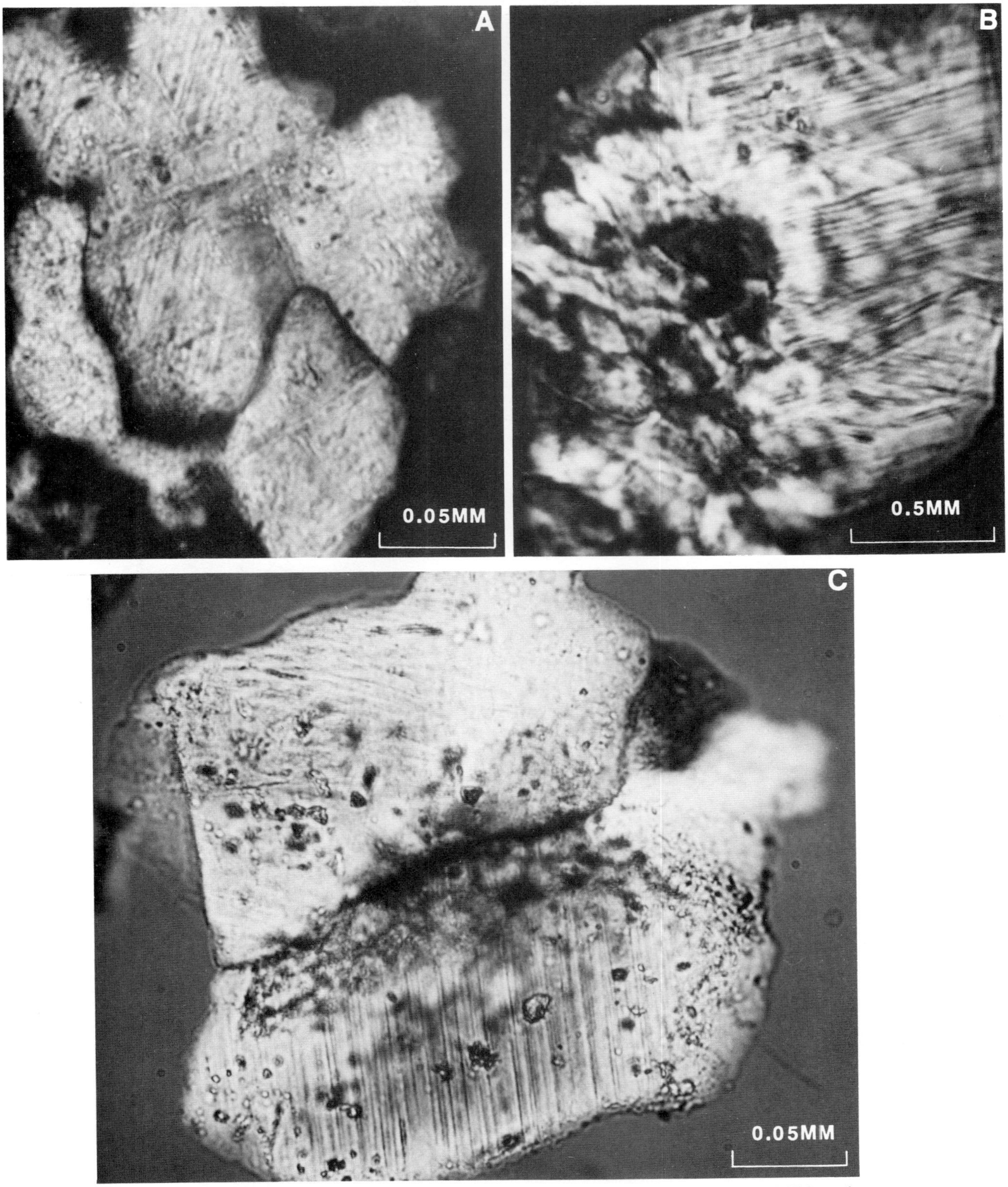

Figure 31. Photomicrographs of shocked quartzite or metaquartzite? grains from the K/T boundary impact layer. A, quartzite or metaquartzite? grain from the Berwind Canyon K/T boundary site, about 20 km northwest of Trinidad, Colorado. B, quartzite grain from the Brownie Butte, Montana, locality composed of one large quartz optical unit and numerous smaller optical units. C, quartzite grain from the Clear Creek North K/T boundary site, about 10 km south of Trinidad, Colorado, composed of several optical units. Photographed with cross-polarized light.

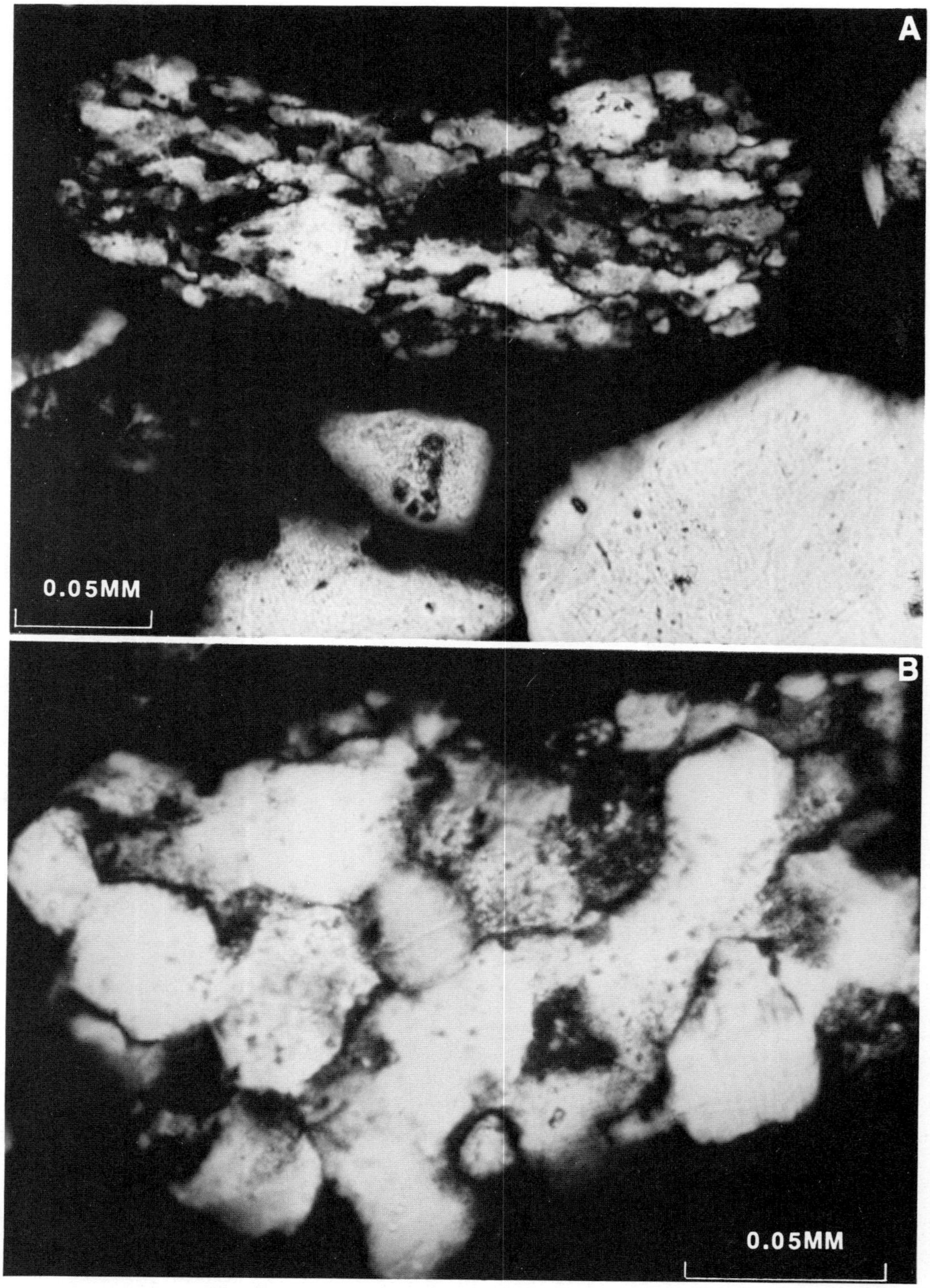

Figure 32. Photomicrographs of shocked metaquartzite and quartzite? grains. A, metaquartzite grain from the Berwind Canyon K/T boundary site, about 20 km northwest of Trinidad, Colorado, showing foliated nature of grain. B, metaquartzite or quartzite? grain. Under high magnification, multiple sets of shock lamellae can be seen in optical units of each grain. Photographed with cross-polarized light.

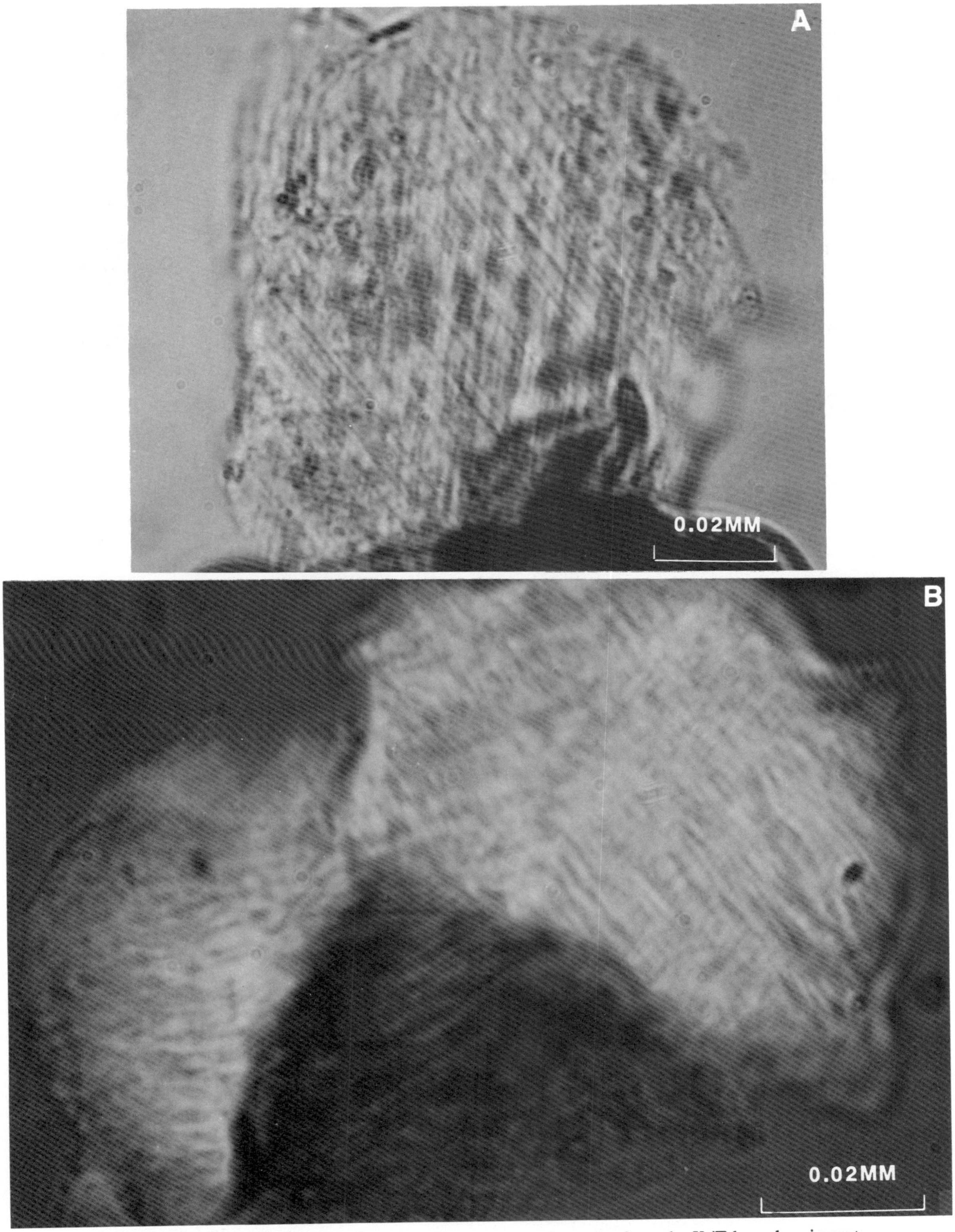

Figure 33. Photomicrographs of shocked quartz and quartzite grains from the K/T boundary impact layer at Stevns Klint, Denmark. A, quartz grain showing two sets of shock lamellae. B, quartzite grain contains three optical units, each optical unit contains multiple sets of shock lamellae. Photographed with cross-polarized light.

Figure 34. Photomicrograph of a double polished thin section of clastic mineral grains from below the K/T boundary claystone. Grains from carbonaceous shale a few millimeters thick that directly underlies the boundary claystone at the Berwind Canyon K/T boundary site, about 20 km northwest of Trinidad, Colorado. Assemblage includes mostly quartz and a small number of quartzite and chert grains, none of which contain shock features. Photographed with cross-polarized light.

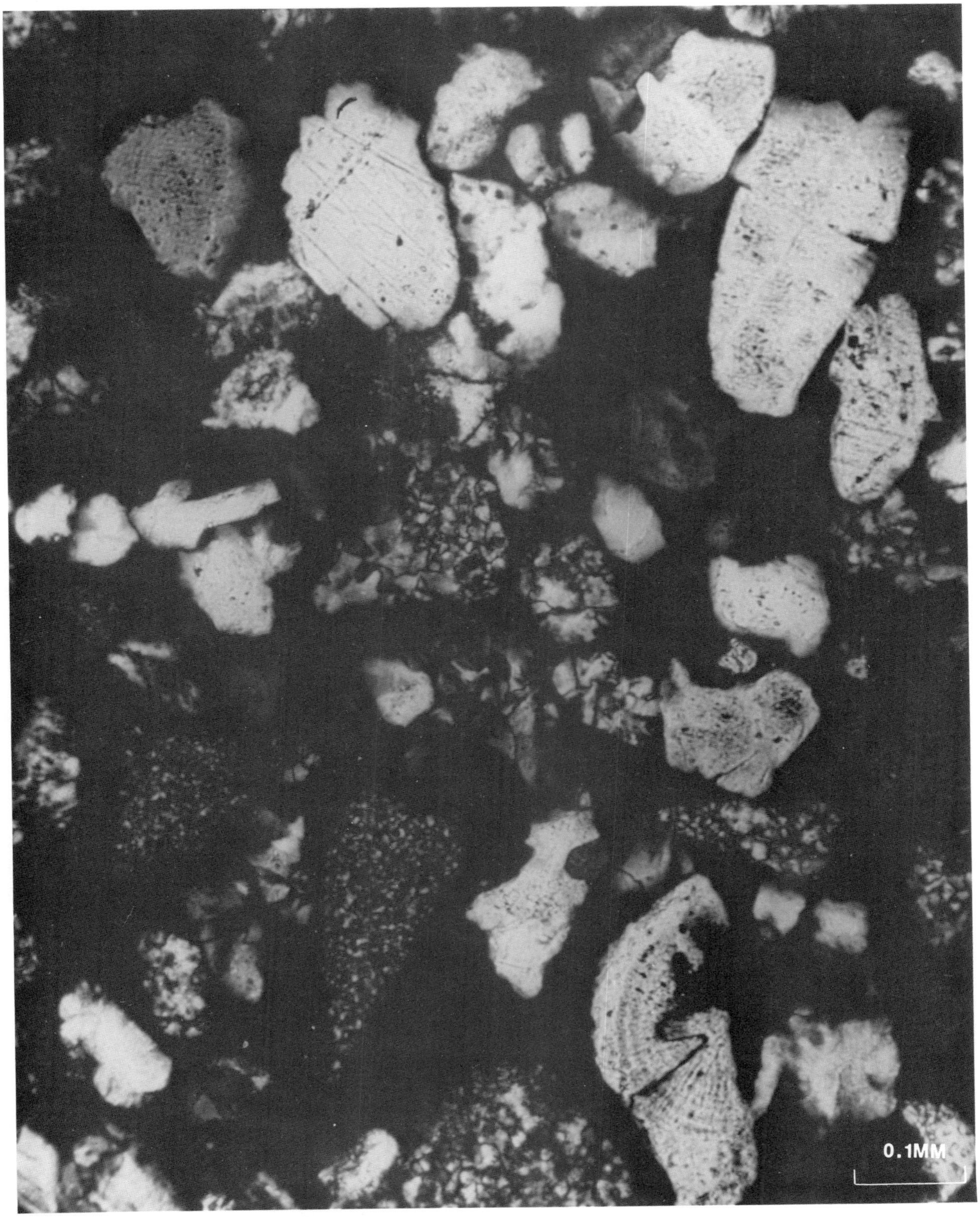

Figure 35. Photomicrograph of a doubly polished thin section of clastic mineral grains from the K/T impact layer. Grains from the same site (Berwind Canyon) as those in Figure 34. Several shocked quartz grains are in upper part of photomicrograph. Assemblage of grains is markedly different than those shown in Figure 34. Notice the high percentage of quartzite, metaquartzite, and chert grains. Unusual grain showing circular-radial structure in lower right corner of photomicrograph. Photographed with cross-polarized light.

SEM showed are iron sulfide. Reflected-light microscopic study of the cubic sulfide grains shows that they are isotropic; therefore, they are pyrite rather than marcasite.

The occurrence of chalcedony grains in the impact layer in the Raton basin and their apparent absence in underlying rocks argue for an unusual event to explain their abrupt appearance (compare Tables 1 and 9 and Figs. 34 and 35). The abrupt appearance of chalcedony grains containing tiny cubic pyrite crystals at each K/T site in the Raton basin indicates that its provenance may not have been from the normal erosion of older rocks adjoining the basin. Shock lamellae were not seen in any of the chert or chalcedony grains in the impact layer.

An exceedingly rare type of shock-metamorphosed lithic fragment composed of an aggregate of sodic plagioclase, potassium-rich feldspar (microcline), and quartz occurs in the impact layer in the Raton basin. Shock lamellae occur in the quartz and microcline of these grains (Fig. 36). This composite-type shock-metamorphosed grain suggests that the K/T impact target rock included a small amount of granitic rock.

Plagioclase and a potassium-rich feldspar compose less than a few percent of the grains in the impact layer (Table 9, Fig. 37) at most K/T sites. Considerable feldspar occurs in the impact layer at the Madrid East site as noted in a foregoing paragraph, but shock lamellae do not occur in these grains. Optical properties of the shocked plagioclase grains, including their indices of refraction, optic angles, cleavage planes, extinction angles, and shock lamellae orientations, were determined in the immersion cell of a Wilcox spindle stage, and the orientations of planar features were plotted on a stereonet. The optical data indicate that the plagioclase grains are mainly oligoclase. Potassium feldspar grains are low in sodium, as judged by their refractive indices and by electron-microprobe analyses. Some potassium-rich feldspar grains show typical albite-pericline grid twinning, which indicates that they are triclinic and surely microcline. A few of the potassium feldspar grains studied contain several sets of shock lamellae, and others show shock mosaicism. Another indication that some potassium feldspar grains are shock-metamorphosed is that their refractive indices and optic angle are slightly lowered below their lowest normal values. The optic angle ($2V_x$) of one potassium feldspar (microcline) grain was about 30°. The general composition of the feldspar grains (oligoclase and microcline) suggests that they were derived from silicic, rather than basic, rocks at the impact site.

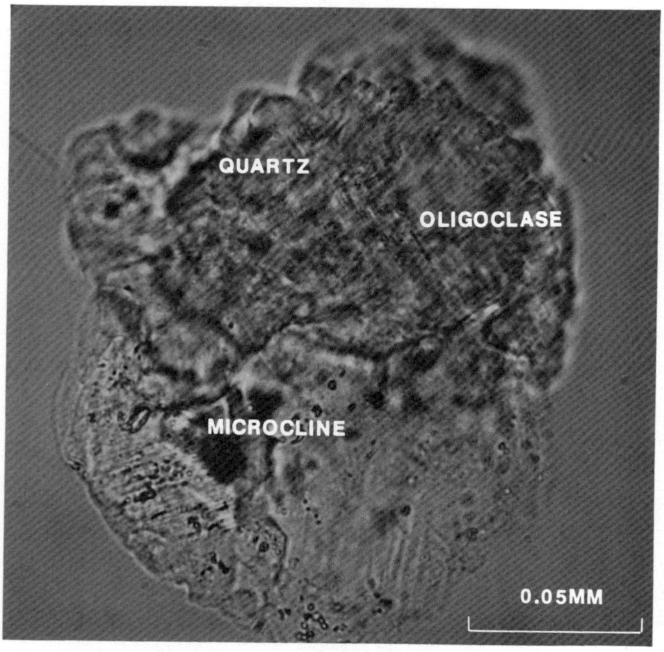

Figure 36. Photomicrograph of a shocked granite-like grain. Such grains composed of quartz, oligoclase, and microcline are rare and indicate that the rocks at the K/T boundary impact site were, in part, granitic. Faint shock lamellae in quartz and oligoclase can be seen in the grain oriented in this position. Such lamellae are better seen at high magnification when the grain is rotated so lamellae are vertical. Grain is from the K/T boundary impact layer at the Clear Creek North K/T boundary site, about 10 km south of Trinidad, Colorado. Photographed with plane-polarized light.

Bohor and others (1984, p. 867) purportedly observed about 2 percent sanidine in the insoluble residue of the K/T boundary claystone and impact layer at Brownie Butte, Montana. However, they did not present either optical microscopic or x-ray diffraction data supporting their claim. Perhaps they identified the potassium feldspar using only an SEM and did not check its optic and x-ray properties. Potassium feldspar grains from the Brownie Butte K/T boundary impact layer that I studied optically consisted almost entirely of microcline and not sanidine. Exceedingly rare sanidine grains were identified in a few samples of K/T boundary rocks, but rare sanidine grains can be found in most Cretaceous rocks of the Western Interior of North America.

In summary, several different mineralogical types of clastic mineral grains in the K/T boundary impact layer show shock-metamorphic features and are thought to reflect the composition of target rocks at the K/T impact site (Fig. 38). Judging from the amount of shock-metamorphosed quartzite, metaquartzite, and quartz in the impact layer, the target rocks were probably sedimentary, metasedimentary, and granitic rocks. A minuscule amount of shock-metamorphosed oligoclase, microcline, and granite-like mixtures of feldspar and quartz suggest that the target rock was, in part, granitic.

Many grains (quartzite, metaquartzite, and chert) in the K/T boundary impact layer do not show shock-metamorphic effects. The source rocks for these mineral types is clearly different than the source rocks for subjacent sedimentary rock units. For example, lithic grains of quartzite, metaquartzite, and chert generally constitute only a few percent of the grains in the carbonaceous shale, mudstone, and sandstone that lie below the K/T boundary claystone (Fig. 34 and Table 1). Yet, in the impact layer, these grains are a significant component of the assemblage of grains (Figs. 35 and 38; Table 9). This remarkable change takes place at the same stratigraphic position at all K/T boundary sites in the Raton basin and at boundary sites in Montana and Wyoming and suggests that most, but not all, of these unshocked minerals are impact derived.

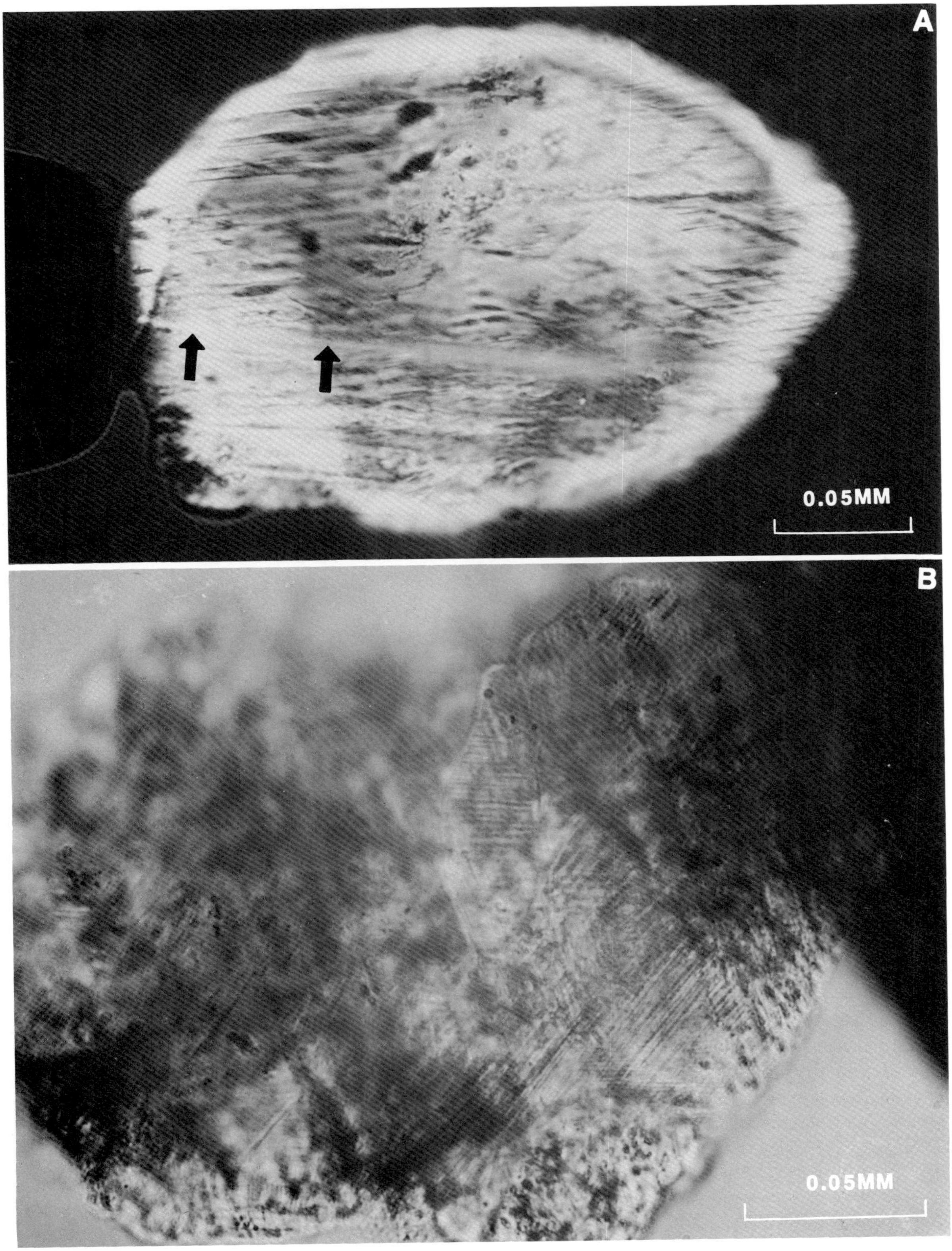

Figure 37. Photomicrograph of shocked feldspar grains from the K/T boundary impact layer. A, oligoclase grain from the Brownie Butte, Montana, locality showing two prominent sets of shock lamellae. Albite twin lamellae shown at arrow. Photographed with cross-polarized light. B, microcline grain from the Brownie Butte, Montana, locality showing two prominent sets of shock lamellae. Photographed with plane-polarized light.

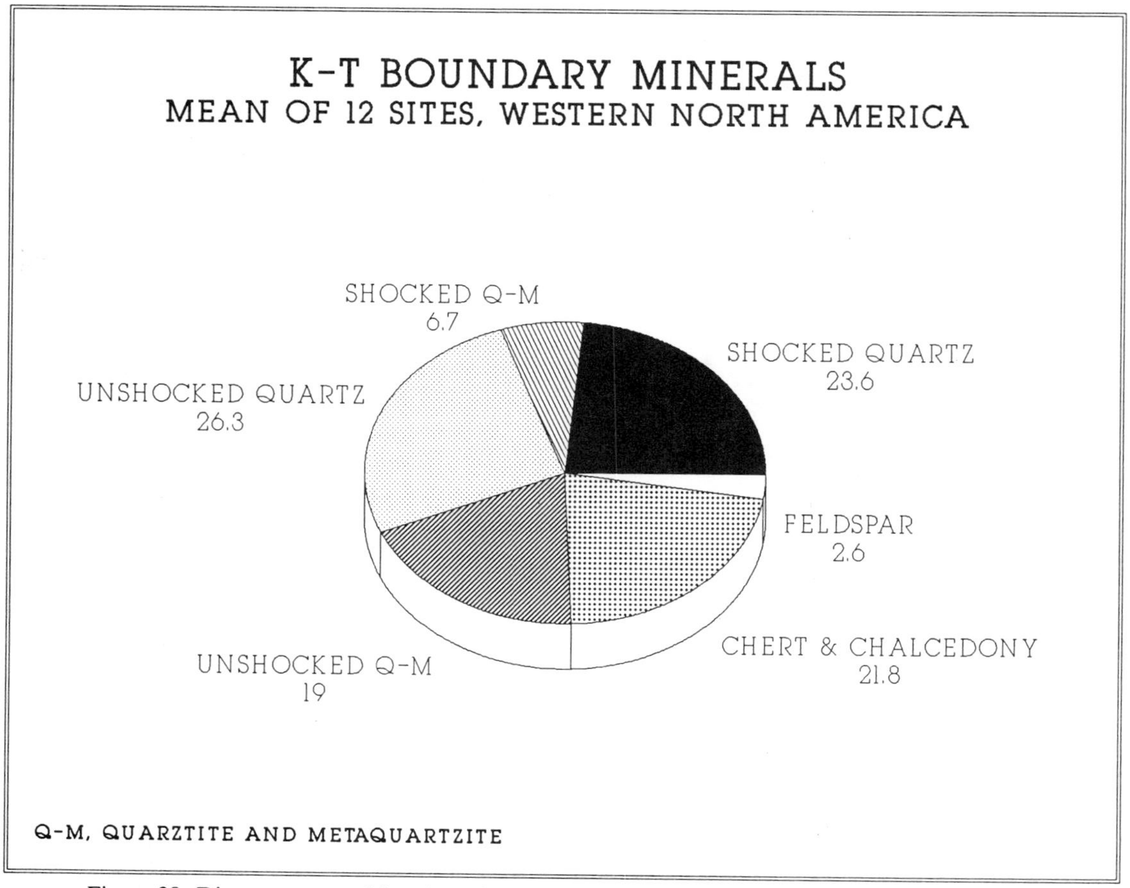

Figure 38. Diagram summarizing the chief mineralogic types of clastic mineral grains in the K/T boundary impact layer from 12 sites in western North America. Thirty percent of the grains are shock metamorphosed. Data tabulated in Table 9.

SHOCK-METAMORPHIC FEATURES IN K/T BOUNDARY CLASTIC MINERALS

Shock lamellae

Shock-metamorphic features in quartz grains of K/T boundary rocks were recognized first by Bohor and others (1984). Their discovery provided important physical evidence confirming that an asteroid struck Earth at the close of the Cretaceous Period as suggested by Alvarez and others (1980). The effects of shock-metamorphism in K/T boundary rocks consisted of multiple intersecting sets of planar microstructures in quartz grains. These microstructures have been termed "shock lamellae" by Chao (1967), "planar features" by Carter (1965), "planar elements" by Stöffler (1972), and "microfractures" by Chao and Goresy (1973, p. 291). The planar microstructures can be seen readily using a petrographic microscope and plane-polarized light, and these microstructures give the quartz grains a remarkable appearance (Figs. 30 and 39) that is far different from normal unshocked quartz. In particular, quartz grains with multiple sets of planat microstructures have a strikingly different appearance than quartz of tectonites. Such deformed quartz grains containing diffuse curved lamellae can be found in most backyard sand piles. Those petrographers unfamiliar with the appearance of shock-metamorphosed quartz might be hesitant to sight-identify the mineral as quartz without determining its optical and x-ray properties, because of its highly unusual appearance. For those who have studied shock-metamorphosed rocks, the conclusion that the multiple sets of planar lamellae in K/T boundary minerals are of shock-metamorphic origin is unambiguous.

The microscopic properties of shock lamellae are best seen when they are oriented nearly parallel to the axis of a petrograhic microscope. In this orientation, the lamellae appear as parallel bright lines commonly bordered by dark lines. The shock lamellae differ from twin lamellae in plagioclase feldspar in that adjacent shock lamellae do not go to extinction during rotation of the stage of a petrographic microscope as do twin lamellae in plagioclase. Transmission electron microscopic (TEM) studies have shown that shock lamellae in minerals that have experienced low shock are generally planar open microfractures, dislocation bands, microgranulated laminae, and tiny voids (Xie and Chao, 1985, p. 108, and Figs. 4 and 6). Shock lamellae in minerals that have experienced moderate to high shock consist of submicro-

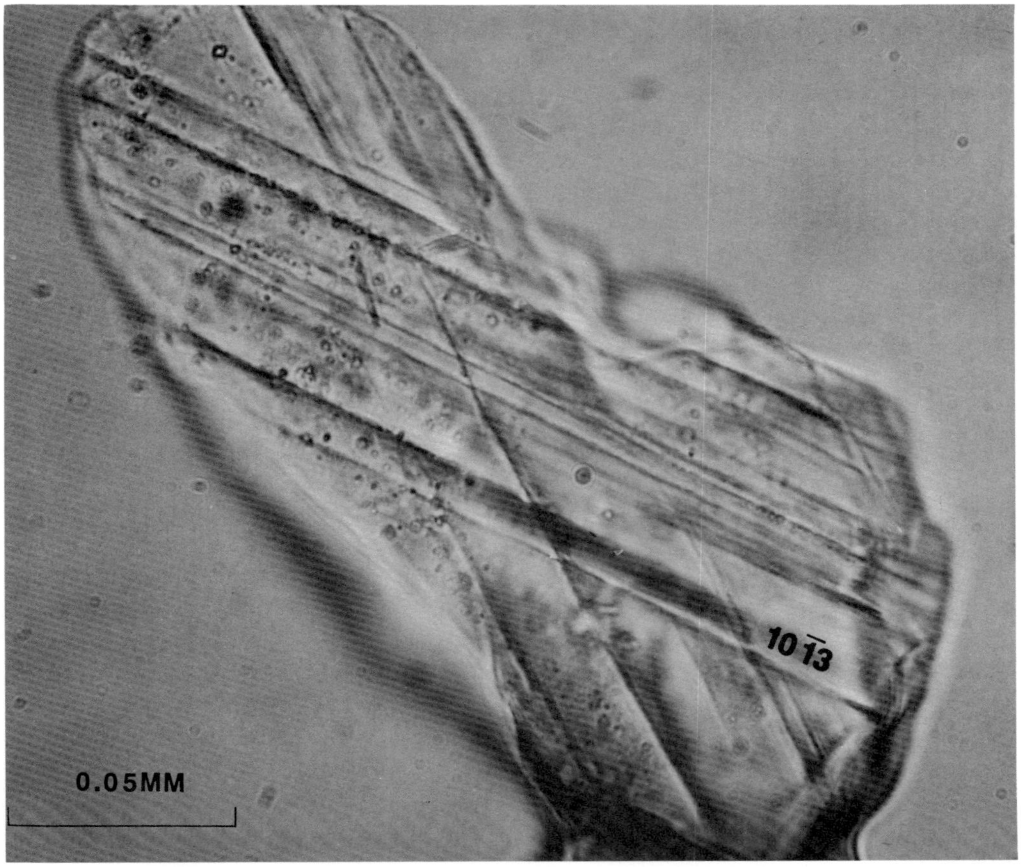

Figure 39. Photomicrograph of a shocked quartz grain. Two prominent sets of shock lamellae are partly defined by fluid inclusions. Such sets have been termed "decorated lamellae" by Carter and Friedman (1965) and Robertson and others (1968). Grain from the K/T boundary impact layer at the Clear Creek North K/T boundary site, about 10 km south of Trinidad, Colorado. Photographed with plane-polarized light.

meter-thick lamellae of silica glass (Chao and Goresy, 1973, Fig. 3; Xie and Chao, 1985, p. 108, and Fig. 7).

Optical measurement of shock-lamellae orientations in quartz grains shows that they parallel rational crystallographic planes, and the most common orientation is about 23 degrees from the basal plane and parallel to Omega, the form (10$\bar{1}$3). Figure 40 is a frequency diagram showing the orientation relative to the basal plane of 200 sets of shock lamellae in quartz grains from the Clear Creek North, Colorado, site (Fig. 2). About 50 percent of the total number of measured sets in quartz are oriented parallel to the form Omega (10$\bar{1}$3). In quartz grains that have only two or three lamellae sets, the planes are more widely spaced than in grains having a large number of differently oriented sets. Although most lamellae are planar, some are gently curved (Fig. 41). Lamellae are more likely to be curved in grains having widely spaced sets. Spindle stage study of 180 randomly chosen shocked quartz grains from eight K/T boundary sites in the western United States showed that, on average, the grains contain 2.8 planar sets per grain. However, many grains only contain one set of planar microstructures.

Some shocked quartz grains contain numerous relatively large inclusions, a few micrometers in diameter, that under low magnification of a petrographic microscope appear to be opaque (Fig. 30C). Many of these relatively large inclusions, which have highly irregular shapes, are arranged along poorly defined planes. Under high magnification (1,250×), some of these large inclusions are seen to be translucent and essentially colorless. The colorless inclusions have refractive indices that are much less than the host quartz. The exact nature of these inclusions is unknown, but some may be voids and others may be filled with fluid. Such inclusions are unlike the globular glass inclusions commonly found in silicate phenocrysts of pyroclastic volcanic rocks (Anderson, 1976; Anderson and others, 1989). In addition to these inclusions, rare shocked quartz grains contain minute zircon and biotite crystals.

In some shocked quartz grains, submicrometer-size inclusions define or parallel the planes of the shock lamellae. These inclusions can be seen at high magnification (1,250×) using an oil-immersion objective of a petrographic microscope (Fig. 39). Most small inclusions have globular shapes, but some are highly

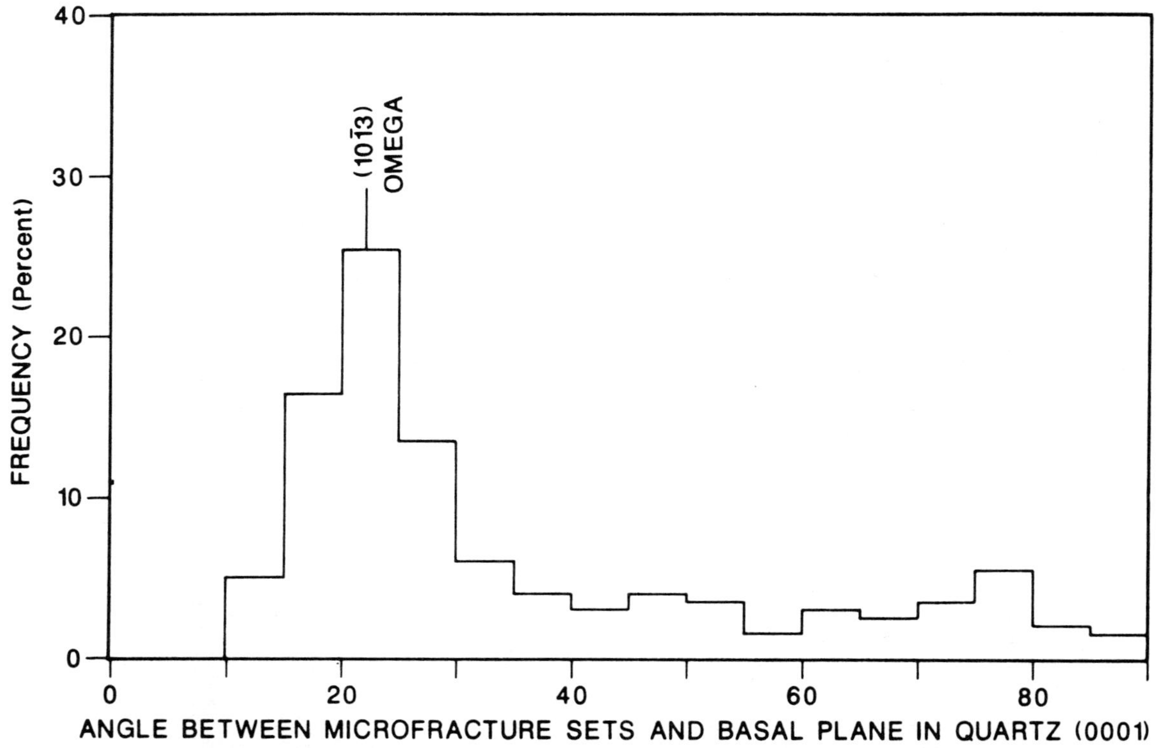

Figure 40. Frequency diagram showing the orientation of planar shock lamellae in quartz grains. Orientation of 200 planar sets measured in grains from the K/T boundary impact layer from the Clear Creek North K/T boundary site, about 10 km south of Trinidad, Colorado. About 50 percent of the shock lamellae are oriented about 23 degrees from the basal plane and parallel to the form Omega (10$\bar{1}$3).

irregular in outline. Some inclusions are thought to be filled with a fluid because they contain a tiny gas bubble that moves with Brownian motion. Such fluid inclusions oriented like a string of beads along the planar features were termed "decorated lamellae" by Carter and Friedman (1965) and Robertson and others (1968). In grains having individual shock lamellae that are widely spaced, the fluid inclusions are relatively large. In grains having closely spaced, fine lamellae, the fluid inclusions are relatively small and can barely be discerned. In many shocked quartz grains, the presence of fluid inclusions along the lamellae cannot be confirmed because their size may be below the resolving power of a petrographic microscope. In short, a complete gradation probably exists in fluid inclusion size from those that can be seen at high magnification (1,250×) to those that cannot be resolved with an optical microscopic.

The origin of fluid inclusions oriented along shock lamellae or so-called "decorated lamellae" in K/T boundary minerals and in shock-metamorphosed minerals at impact structures is problematic. However, such inclusions in minerals at impact sites may originally have been voids, subsequently filled with fluid, that formed along planar microfractures created by tensional release following the passage of a shock wave. Following the release of pressure, some microfractures closed tightly, but others closed imperfectly and left voids along sections of the microfractures (see Xie and Chao, 1985, p. 108).

Stöffler (1972) attributed the origin of fluid inclusions in shocked quartz grains to post-shock annealing in a high-temperature environment. This interpretation for the origin of the fluid inclusions in shock-metamorphosed K/T boundary minerals is inconsistent with the fact that K/T boundary minerals did not undergo high-temperature annealing at the impact site. Instead, shock-metamorphosed minerals were lofted instantaneously into the atmosphere, quenched, and transported to their ultimate depositional sites without undergoing high-temperature annealing.

Indices of refraction of shocked quartz

Another property of quartz and other silicate minerals that have experienced hypervelocity shock is the lowering of their indices of refraction and birefringence. In contrast, the refractive indices of quartz in tectonites are never lowered below their normal values ($n_o = 1.544$; $n_e = 1.553$) according to Chao (1967). TEM studies have shown that slight lowering of the indices of refraction of shocked quartz is related to the presence along shock lamellae of dislocation bands, microfractures, and voids (Xie and Chao, 1985, p. 109). Marked lowering of the indices of refraction of quartz in moderately to strongly shocked rocks was related to the presence of silica glass along shock lamellae. For example, Xie and Chao (1985) showed that the mean index of refraction of

Figure 41. Photomicrograph of a shocked quartz grain. Three prominent sets of slightly curved shock lamellae can be seen in this orientation of the grain. Notice large opaque inclusions of unknown composition in grain. Grain from the K/T boundary impact layer at the Clear Creek North K/T boundary site, about 10 km south of Trinidad, Colorado. Photographed with plane-polarized light.

most shocked quartz in rocks at the Ries impact structure in Germany had been lowered from n = 1.5485 to n = 1.46.

Some shocked quartz grains in the K/T boundary impact layer in western North America have indices of refraction slightly below their normal values. P. J. Modreski of the U.S. Geological Survey found that the indices of refraction of shocked quartz in K/T boundary rocks in Montana are lowered slightly below their normal values (reported in Bohor and others, 1984). Similarly, in the Raton basin, the refractive indices of shocked quartz in the impact layer are lowered, but not drastically. Figure 42 shows that only about 30 percent of shock-metamorphosed quartz grains in the impact layer in the Raton basin have had their refractive indices lowered, but even in these grains the indices of refraction have not been lowered below n_o = 1.535 indicating that the quartz grains were only weakly shocked. This result is curious in view of the fact that the indices of refraction of shock-metamorphosed quartz in rocks at moderately large impact structures such as the Ries (25 km in diameter) in southern Germany have been drastically lowered, as low as quartz glass (n = 1.46). The K/T boundary asteroid is thought by some (Alvarez and others, 1980) to be much larger than the one that formed the Ries structure, and accordingly, one might expect to find marked lowering of the refractive indices and birefringence of quartz in the K/T boundary impact layer. However, the size of the asteroid is not the only factor that controls the degree of shock-metamorphism such as the lowering of the refractive indices of some quartz grains in target rocks. The nature of the target rocks is also critical. For example, the K/T impact may have occurred in an area such as Manson, Iowa, underlain by a thick sequence of sedimentary and metasedimentary rocks that may have responded differently to the shock waves than the crystalline metamorphic and granitic rocks of the Ries crater area, which were buried by only a thin sedimentary cover.

Mosaicism

Yet another optical property of silicate minerals that have undergone hypervelocity shock at impact structures and nuclear explosion sites has been termed "shock mosaicism" by Stöffler (1972, p. 87). The optical effects include a highly irregular patchy extinction pattern probably involving the formation of disoriented domains in the crystal structure (Fig. 43). Shock mosaicism

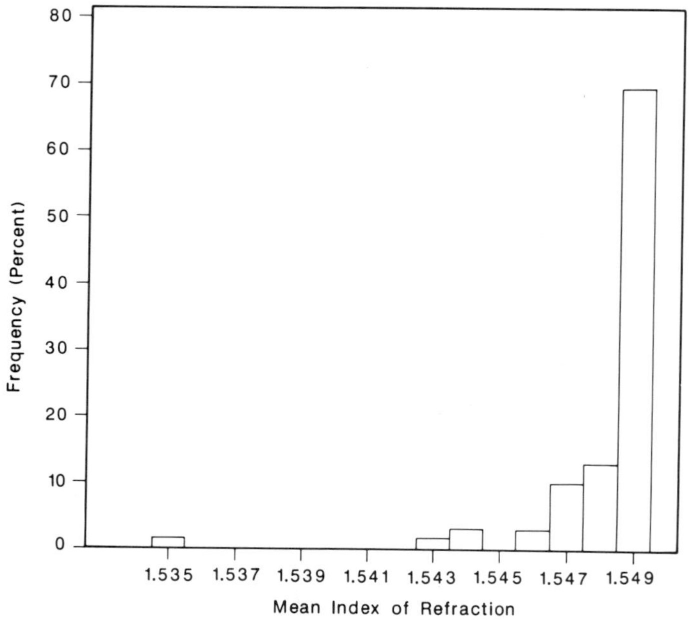

Figure 42. Frequency diagram showing the average index of refraction of shocked quartz grains. Indices of refraction of 63 quartz grains from the K/T boundary impact layer at the Clear Creek North, Colorado, locality were measured. Indices of refraction measured using the focal-masking technique described by Wilcox (1983).

commonly occurs in quartz and feldspar of the K/T boundary impact layer, not only in grains containing shock lamellae but also in some grains without such features.

Thetomorphic glass

Quartz grains in rocks at large impact sites that have experienced moderate to high levels of shock typically contain vitreous material along shock lamellae. This type of glass was termed "thetomorphic glass" by Chao (1967), and it forms by shock-induced solid-state conversion of a mineral such as quartz to silica glass, not by thermal melting of the quartz. He (1967, p. 195) pointed out that if the glass had formed by melting of the quartz it should contain vesicles and flow structure. Thetomorphic silica glass is common in moderately to highly shocked quartz in rocks at the Ries impact structure (Chao, 1967; Stöffler, 1971b; Xie and Chao, 1985). Thus, thetomorphic silica might be expected to occur along planar features in some shock-metamorphosed quartz grains in K/T boundary rocks. This expectation is based on the assumption that the K/T boundary asteroid was larger than the one that formed the Ries impact structure, and one, therefore, would expect that some of the K/T impact target rocks would have experienced moderate to high levels of shock.

No trace of thetomorphic glass was recognized by optical microscopic methods in any shocked quartz grains from the Raton basin, particularly those quartz grains that contain the most numerous sets of shock lamellae. The fact that thetomorphic glass was not identified optically in shocked quartz grains is not surprising because the indices of refraction of only about 30 percent of the shocked quartz grains are lowered below their normal values (Fig. 42). Moreover, the indices of refraction of these quartz grains are slightly lowered, which indicates that they contain only a trace amount of thetomorphic silica glass, if any. Although thetomorphic silica glass was not recognized by optical microscopic methods in shocked quartz grains from the Raton basin, Gratz and others (1988) claimed to have found thin "glass layers" along shock lamellae in a few quartz grains from the Clear Creek North site. They used TEM methods to identify glass lamellae in the shocked quartz grains.

One explanation for the lack of significant amounts of thetomorphic silica glass in shocked quartz of K/T boundary minerals is that the K/T impact probably occurred in an area such as Manson, Iowa, that is underlain by about 0.7 km of Phanerozoic and 6 km of Proterozoic sedimentary rocks. These rocks probably responded to the K/T boundary impact shock wave in a different way as compared to the shock wave that affected the crystalline metamorphic and granitic rocks in the Ries crater area. The apparent lack of significant amounts of shock-produced thetomorphic glass in quartz grains of the K/T boundary impact layer suggests that stishovite and coesite, high-pressure (>30 kb) polymorphs of SiO_2, also may not be found. Thetomorphic glass forms at lower pressures than does either coesite or stishovite (Chao, 1967).

Stishovite

Bohor and others (1984) reported that they found traces of stishovite, but oddly not thetomorphic silica glass or coesite, in mineral concentrates obtained from the K/T boundary claystone (not the impact layer) at Brownie Butte, Montana. They claimed to have seen and measured a few faint x-ray diffraction lines of a mineral they imagined to be shishovite on x-ray films of a single shocked quartz grain and a group of ten shocked quartz grains. They further claimed that stishovite lines were seen on x-ray films of a hydrofluoric-acid insoluble residue of the K/T boundary claystone at the same site. The acid-insoluble residue they used was small, only a few milligrams. The alleged presence of stishovite in quartz grains with multiple sets of planar lamellae was used to support their conclusion that the sets of planar lamellae were of shock-metamorphic origin. E. E. Foord of the U.S. Geological Sruvey, who performed the x-ray analysis, kindly made an x-ray film of the alleged stishovite available to me. B. F. Leonard of the U.S. Geological Survey and I examined and remeasured the lines on the films, and contrary to the report of Bohor and others (1984), we did not find any lines that would indicate the presence of stishovite in the x-ray films they gave me.

I find it difficult to accept the evidence that Bohor and others (1984) presented in support of their claim that stishovite occurs in K/T boundary rocks for the following reasons. They used single shocked quartz grains and such small concentrates of K/T boundary minerals that it seems highly unlikely that they could have detected stishovite using the x-ray diffraction method. The detection limit for stishovite, set in the following paragraphs

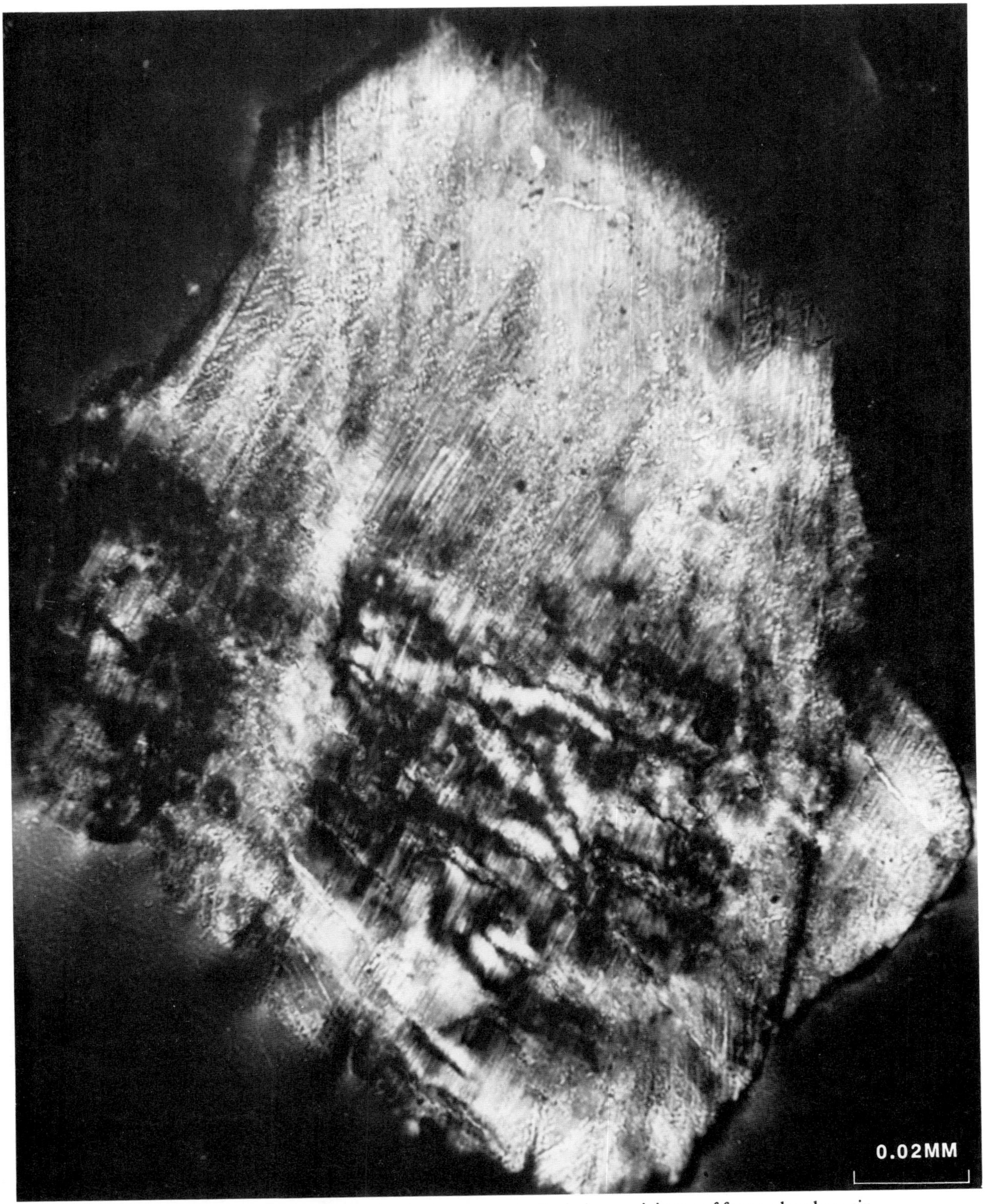

Figure 43. Photomicrograph of a shocked quartz grain. Grain has a minimum of four and perhaps six sets of shock lamellae in this orientation. An area of anomalous extinction termed shock "mosaicism" occurs in the lower part of the grain. Grain from the K/T boundary impact layer at the Clear Creek North K/T boundary site, about 10 km south of Trinidad, Colorado. Photographed with cross-polarized light.

for K/T boundary shocked mineral concentrates, is far too low for it to be detected by x-ray diffraction methods in the single grains and small samples that Bohor and others (1984) used.

McHone and others (1989) also reported finding trace amounts of stishovite, but strangely not coesite, in splits of an 8-g mineral concentrate of K/T boundary minerals from the Raton, New Mexico, site (Fig. 2). They used two different methods in their search for stishovite: magic angle spinning nuclear magnetic resonance spectroscopy (MASNMR), and powder x-ray diffraction. The MASNMR method takes advantage of the fact that the silicon in stishovite is in octahedral coordination with oxygen rather than the normal tetrahedral coordination of silicates. Trace amounts of stishovite can be identified in shocked quartz grains using this method (Smith and Blackwell, 1983; Yang and others, 1986).

Three boundary interval samples were collected at the Raton site by McHone and others (1989), and each, according to them, consisted of the boundary claystone and as much as 2.0 cm of overlying "coal-rich clays." My study of the K/T boundary interval at this site (Fig. 15) showed that they probably collected at the westernmost of the two outcrops and that their samples actually consisted of three units: (1) the K/T boundary claystone, about 1.5 cm thick, which essentially does not contain quartz grains; (2) the K/T boundary impact layer, about 0.3 cm thick, which always contains a concentration of shocked quartz grains; and (3) gray mudstone, 1.0 to 6.0 cm thick, which contains about 0.01 to 1.0 percent unshocked quartz and rare shocked quartz grains. Thus, the gray mudstone that overlies the impact layer probably contributed a significant amount of unshocked quartz to the residue that they argued contained stishovite.

McHone and others (1989) prepared three nonclay mineral concentrates from the three samples (each 8 kg) of the boundary interval at the Raton site. Each mineral concentrate weighed about 8 g. Unfortunately they did not describe the mineralogic characteristics of these concentrates before they pulverized them. In particular, they did not estimate the amount of shocked quartz grains relative to unshocked quartz grains in their samples, and the lack of this data seriously compromises their results. As stressed in the previous paragraph, they may have inadvertently included a considerable amount of material containing unshocked quartz grains in their three primary samples. Nonetheless, they identified stishovite in all splits, each about 0.5 g, separated from *only one* of their 8-g concentrates using MASNMR spectroscopy.

Stishovite was also identified by powder x-ray diffraction methods in a hydrofluoric-acid insoluble residue (sample size not specified) of one of the splits used for MASNMR analysis (McHone and others, 1989). They assigned eight x-ray diffraction peaks to shishovite on the diffractogram of the residue (see their Fig. 2). The strongest peak for stishovite on their diffractogram is amazingly large considering the low concentration (2 to 3 mg) of this mineral in the analyzed split. McHone and others (1989) did not mention the fact that stishovite and rutile are isostructural; that is, both are octahedral. Accordingly, they have nearly identical x-ray diffraction patterns. Rutile is a relatively common mineral in K/T boundary mineral concentrates. Moreover, the strongest x-ray diffraction line of stishovite (30.18° 2θ) is nearly identical to the strongest x-ray diffraction line of goyazite (30.17° 2θ), a relatively common diagenetic mineral in K/T boundary mineral concentrates that are not completely digested in concentrated HNO_3.

I also searched for stishovite in K/T boundary minerals from a site that contains one of the highest concentrations of shocked quartz grains in the Raton basin: the Clear Creek North site in Colorado (Fig. 2, Table 9). Two attempts were made to find stishovite in different acid-insoluble mineral-grain concentrates from this site. These concentrates were separated from the impact-layer claystone as described in a previous section of this paper. The first attempt was made on the acid-insoluble residue (7.780 g) obtained from a 3-kg sample of the impact layer (here called concentrate A), and the second attempt was made on a larger acid-insoluble residue (16.867 g) obtained from a 10-kg sample of the impact layer (here called concentrate B). Such residues, as previously described, consist primarily of mineralogic varieties of SiO_2—quartz, quartzite, metaquartzite, chert, and chalcedony—but also trace amounts of zircon and authigenic rutile, barite, and gypsum. Stishovite should only occur in about 30 percent, on average, of the mineral grains (Table 9), only those quartz grains that contain shock lamellae.

The mineral grains of concentrate A were digested in a mixture of hydrofluoric (48 percent) and hydrochloric acids in a new 250-ml teflon dish following the general procedure outlined by Fahey (1964) for the recovery of stishovite. The procedure of Fahey (1964) was modified slightly in that the hydrofluoric-hydrochloric acid insoluble residuum from concentrate A was further digested in nitric and sulfuric acids to remove authigenic gypsum and goyazite that pervasively occur in small amounts in K/T boundary rocks. The mineral-grain concentrate thus obtained consisted of several milligrams of material, and optical study indicated that the material was composed chiefly of small (20 to 50 μm) zircon, rutile, and other heavy detrital mineral grains and a miniscule amount of a cubic iron sulfide mineral. The fact that the zircon grains are all individual rounded grains and not compound grains suggests that they were of detrital rather than impact origin. The rutile and iron sulfide are clearly authigenic. EDS-SEM analysis indicated that the chief mineral grains in the residue consisted of 66 percent zircon, 33 percent rutile, and a small amount of cubic iron sulfide. One grain of a chromium-rich spinel was identified using the EDS analyzer of an SEM. X-ray diffraction analysis confirmed the presence of zircon, rutile, pyrite, and possibly marcasite. No lines of stishovite were seen on the powder diffractogram pattern of the residue (detection limit probably 5 percent).

A second attempt was made to find stishovite in a mineral grain concentrate (concentrate B) that was about twice as large as concentrate A. However, the procedure used for concentrate B was modified from that used for concentrate A in that most of the detrital zircon, authigenic rutile, and other heavy minerals first were removed using heavy liquid (2.86 specific gravity) separation techniques. The heavy minerals, including zircon and rutile, were removed from concentrate B prior to acid digestion so that they would not add unwanted bulk to the final residuum that theoretically would contain stishovite. Only 0.094 g of material heavier than 2.86 specific gravity was recovered from concentrate B, and optical microscopic, powder x-ray diffraction, and SEM-EDS analyses confirmed that the heavy minerals again consisted of zircon, rutile, and goyazite. The material lighter than 2.86 specific gravity (16.773 g) was digested using the modified Fahey (1964) procedure as discussed in the previous paragraph. The concentrate thus obtained weighed about 0.0005 g, and optical microscopic, EDS-SEM, and powder x-ray diffraction analyses showed that the residue, theoretically lighter than 2.86 specific gravity, still contained a miniscule amount of heavy minerals, including rutile (38 percent), pyrite and possibly marcasite (35 percent), and zircon (27 percent). The mean size of the grains was 27 μm, and their aspect ratio was 1.74. During SEM-EDS analysis, not one grain dominantly composed of SiO_2 (that would suggest the presence of stishovite) was encountered. Moreover, no powder x-ray diffraction peaks for stishovite were seen on the diffractogram of the hydrofluoric-acid insoluble residue.

As indicated by the data presented in the foregoing three paragraphs, it is unlikely that stishovite occurs in amounts that can be detected by x-ray diffraction studies of large-size mineral concentrates from the K/T boundary impact layer in the Raton basin. If stishovite does occur in shocked quartz grains, as the MASNMR spectroscopic data of McHone and others (1989) suggests, it could only occur in extremely small amounts, less than 1 ppm.

The MASNMR data of McHone and others (1989) suggested that stishovite occurs in shocked quartz grains in at least one of the concentrates they analyzed from the Raton site. At face value, their MASNMR evidence seems compelling, but other evidence suggests that stishovite should not occur in significant amounts in shocked quartz grains in K/T boundary rocks. Optical microscopic studies have shown that the principal indices of refraction and birefringence of shocked quartz grains from the impact layer are not dramatically lowered below their normal values (Fig. 42). This implies that the quartz grains may only contain a trace amount of thetomorphic silica glass. Consequently, the quartz grains experienced only low levels of shock pressure insufficient to form coesite or stishovite. Chao (1967) established the paragenetic relations between thetomorphic silica glass and the two high-pressure polymorphs of SiO_2. He (p. 196) stated that ". . . coesite coexists with silica glass in all observed instances." He further commented that "Stishovite occurs naturally in specimens containing a relatively high concentration of coesite. . . ."

In conclusion, much more work needs to be done before the presence of stishovite in K/T boundary shocked quartz grains can be unequivocally accepted. Several questions need to be answered relative to the claim of McHone and others (1989) that trace amounts of stishovite occur in K/T boundary rocks. One question is why did they only find stishovite in one of three samples from the Raton, New Mexico, site? A second question is why did they not provide an estimate of the amount of shocked quartz, the probable host for the stishovite, in their concentrates? Clearly, the results of McHone and others (1989) must be checked to determine if the trace amount of stishovite they detected actually occurs in shocked quartz grains or if tramp laboratory stishovite was accidently incorporated into one of their Raton site samples.

ORIGIN OF SHOCK-METAMORPHIC FEATURES IN K/T BOUNDARY MINERALS

Bohor and others (1984) concluded that the multiple intersecting sets of planar lamellae in quartz in K/T boundary rocks formed by shock-metamorphic processes during the impact of an asteroid that struck Earth at the close of the Cretaceous. They arrived at this conclusion on the basis of two areas of research: (1) the known impact record of the Earth (Shoemaker, 1979; Short and Bunch, 1968; Grieve, 1982); and (2) experimental and empirical mineralogy of impact materials (Wackerle, 1962; Carter and others, 1964; Carter, 1965, 1971; Chao, 1967, 1968; McIntyre, 1968, Ahrens and Rosenberg, 1968; Bunch, 1968; Robertson and others, 1968; Short, 1968; Englehardt and others, 1968; Englehardt and Bertsch, 1969; Stöffler, 1971a, 1972). These studies have shown that the most important evidence for the recognition of shock-metamorphism in silicate minerals is the occurrence of multiple intersecting sets of planar microstructures, succinctly termed "microfractures."

Böhm lamellae in quartz of tectonites might be confused with shock-induced microstructures, but Böhm lamellae are far less regular and generally occur in single sets, rarely in two sets, and never in three or more sets. Bunch (1968, p. 419) stated, "The occurrence of two or more sets of planar features in quartz grains appears to be unique to impact sites." It is important, therefore, that shock-metamorphosed quartz grains in K/T boundary rocks typically average 2.8 planar sets per grain, and some grains have as many as nine sets. Such features have been seen only in silicate minerals that have undergone hypervelocity impact at meteor impact and nuclear and high-yield chemical explosion sites (Chao, 1967). Carter (1968, p. 463) stated, "These planar features are very different from all other microstructures observed in quartz from rocks deformed either naturally or experimentally under static or dynamic loading conditions." Such features apparently can only form at pressures greater than 60 kb (Short, 1968, p. 203).

In 1986, Carter and his colleagues described rare poorly defined microstructures in quartz, plagioclase, and biotite in the Toba Tuff of Sumatra and proposed that their formation was a result of volcanic-induced shock metamorphism. They thus started a debate (Izett and Bohor, 1986b; Carter and Officer, 1986, 1989; Alexopoulos and others, 1987, 1988, 1989) as to whether shock-metamorphic features can form in phenocrysts in a silicic magma or in silicate minerals in wall rocks of magma chambers during explosive pyroclastic eruptions. If the proposal of Carter and his colleagues were correct, it would provide an alternative mechanism for the formation of shock-metamorphic features found in minerals in K/T boundary rocks. As will be shown in a subsequent section of this report, abundant, well-defined planar microstructures in quartz occur in K/T boundary rocks of western North America (Izett, 1987c, 1988) and, to a lesser extent, globally (Bohor and Izett, 1986; Izett, 1987c, 1988; Preisinger and others, 1986; Badjukov and others, 1986). Two sources of information, one from study of K/T boundary minerals and another from study of pyroclastic volcanic rocks, cast doubt on the idea that shock-metamorphic features can form during violent silicic pyroclastic eruptions.

Information gained from study of K/T boundary rocks

Information gained during this study of K/T boundary rocks in the Raton basin and elsewhere in western North America suggests that the shock-metamorphic features in minerals originated during hypervelocity impact of an asteroid and not during large-scale silicic pyroclastic eruptions. Remnants of such imagined large-scale silicic pyroclastic volcanism (large-volume ash-flow sheets) of the proper age (about 66 Ma) in western North America have not been described in the geologic literature.

A seemingly important observation (Izett, 1987c) supporting the contention that shock-metamorphic features in minerals originated during hypervelocity impact of an asteroid and not during large-scale silicic pyroclastic eruptions is that minerals characteristic of silicic volcanic rocks are essentially absent in the K/T boundary claystone and impact layer. The essential lack of such minerals cannot be attributed to their removal during diagenetic alteration of the imagined volcanic precursor of the K/T boundary claystone because some of the more stable minerals, quartz and zircon for example, probably would have survived diagenesis of the alleged volcanic ash precursor. Moreover, the impact layer contains rare nonvolcanic (microcline) feldspar grains. This fact suggests that, had the imagined volcanic precursor contained feldspar, these feldspar phenocrysts also would have survived diagenesis. Significantly, no shock-metamorphosed quartz or feldspar grains in the impact layer were seen to contain glass inclusions that are typically found in silicate phenocrysts of pyroclastic rocks (Anderson, 1976; Watson, 1976; Sommer, 1977; Anderson and others, 1989).

If the discovery of stishovite in K/T boundary rocks in the Raton basin (McHone and others, 1989) withstands scrutiny and can be confirmed, the contention of Carter and others (1986) that the shock lamellae in quartz are of volcanic origin can be dismissed because stishovite has not been found in volcanic rocks.

Of far greater importance relative to the idea that shock-metamorphic features in minerals originated during large-scale silicic pyroclastic eruptions is the fact that about 29 percent, on average, of the shock-metamorphosed grains in the impact layer in the Raton basin are composite and consist of two or more optical units separated by sutured boundaries. These composite grains are, thus, lithic fragments of quartzite and metaquartzite and obviously cannot be primary volcanic phenocrysts. This information is often seemingly ignored by those who favor a volcanic source for the shocked quartz grains; it needs to be emphasized repeatedly.

Information gained from study of pyroclastic rocks

Carter and others (1986) maintained that shock-metamorphosed silicate minerals can be found in silicic pyroclastic rocks such as the Toba Tuff of Sumatra and, by analogy, other large-volume silicic tuffs such as the Bishop Tuff of eastern California. The most important criterion for the recognition of shock metamorphism in silicate minerals is the presence of multiple intersecting sets of planar lamellae (Chao, 1967; Bunch, 1968, p. 419). I stress the fact that Carter and his colleagues (1986) did not find multiple sets of shock lamellae in quartz and feldspar in the Toba Tuff. However, they did describe anomalous extinction in quartz and plagioclase similar to the shock mosaicism mentioned by Stöffler (1972).

Izett and Bohor (1986b) pointed out that the anomalous extinction in rare plagioclase crystals of the Toba Tuff reported by Carter and others (1986) may not be of shock origin. Instead, it simply may be the patchy extinction commonly seen in resorbed and remelted plagioclase xenocrysts in volcanic rocks. This type of patchy or mosaic extinction in plagioclase crystals is frequently recognized and reported by students of volcanic igneous petrology (see Vance, 1965). Recently, Sharpton and Schuraytz (1989) also concluded that the mosaic extinction in rare plagioclase xenocrysts of the Toba is not the result of shock metamorphism. Carter and others (1986) also described diffuse single lamellae in plagioclase feldspar (curiously not sanidine) that they attributed to shock-metamorphic processes.

Carter and others (1986) proposed that silicate minerals in the Toba Tuff do not contain multiple sets of shock lamellae because the lamellae were annealed out during their residence in the high-temperature regime of a cooling ash-flow tuff. If this reasoning were correct, it would account for the lack of such features in quartz and feldspar of silicic pyroclastic rocks. However, shock lamellae are readily found in shock-metamorphosed quartz and feldspar in the high-temperature suevite breccia of the Ries structure in Germany (Chao, 1967, Fig. 1). Grieve (1987, p. 253) stated that impact-generated melt sheets ". . . are heavily charged with shocked and unshocked lithic and mineral clasts (>50 percent) at the top and bottom." Thus, observational evidence indicates that shock lamellae seem to survive high-temperature annealing. A possible explanation for this observation is that some shock lamellae are actually open microfractures, as determined by TEM studies (Chao and Goresy, 1973, Figs. 3 and 5; Chao, 1976, Fig. 7), and it may take complete melting of a grain to erase this type of shock lamellae. Accepting the suggestion of Carter and others (1986) at face value that shock lamellae are annealed out of volcanic rocks, the fact that shock lamellae are common in K/T boundary shocked minerals should confirm the nonvolcanic origin for such grains. If they do not match the composition of the glass-mantled phenocrysts in pumice lumps, they are obviously xenocrysts.

Alternatively, Carter and others (1986) proposed that the reason that silicate minerals of the Toba Tuff do not contain multiple sets of shock lamellae is that such lamellae do not form in the high-temperature regime of silicic magma chambers. This question can be tested by study of phenocrysts in lapilli of air-fall pumice deposits. Pumice lapilli are quenched magma samples, and in air-fall deposits their phenocrysts have not undergone a period of thermal residence as they do in hot ash-flow tuffs. To address this question, I studied polished thin sections of mineral-grain concentrates from air-fall pumice lapilli of the Bishop Tuff of eastern California, the Huckleberry Ridge and Lava Creek Tuffs of Yellowstone National Park, the Bandelier Tuff and Cerro Toledo Rhyolite of the Jemez Mountains of New Mexico, and the Toba Tuff. No anomalous optical features of any kind were seen in the minerals of the pumice lapilli—no undulatory extinction, strain shadows, Böhm lamellae, mosaicism, single sets of shock lamellae, or multiple intersecting sets of shock lamellae. In short, the grains are monotonously uniform and structureless, go to extinction sharply, and only contain small glass inclusions typical of silicate phenocrysts of pyroclastic rocks (Anderson and others, 1989; see also cover photograph of Geology, v. 17, 1989). This evidence confirms the proposal of Carter and others (1986) that shock-metamorphic features do not form in phenocrysts in a magma.

Carter and others (1986) incorrectly assumed that the silicate minerals that contain the alleged shock-metamorphic features are phenocrysts and not minerals accidentally incorporated into the Toba magma or ash flows as they were emplaced. Another test of the idea that the alleged shock-metamorphosed silicate minerals in the Toba Tuff are phenocrysts and not xenocrysts would be to perform electron microprobe analyses on such grains. Hildreth (1977, 1979) showed that phenocrysts in pumice lapilli of the Bishop Tuff, for example, have narrow compositional ranges. Accordingly, analyses of the alleged shock-metamorphosed feldspar grains of the Toba Tuff could be made to determine if their composition matches those of primary plagioclase phenocrysts in pumice samples that are quenched samples of the Toba magma.

The indices of refraction of quartz and feldspar, alleged to contain shock-metamorphic features, in the Toba Tuff, could be measured to determine if they are lower than normal values for these minerals. Lowering of the indices of refraction of quartz is commonly found in shock-metamorphosed quartz (Xie and Chao, 1985), but has never been reported in quartz of tectonites.

Two obvious choices are available to address the question of origin of the xenoliths and xenocrysts in ash-flow tuffs that potentially could contain shock-metamorphic features. The first choice is that the xenoliths and xenocrysts are picked up from the land surface by an ash flow during its turbulent emplacement. The fact that ash-flow tuffs contain xenocrysts is well known. Ross and Smith (1961, p. 35) stated, "A great variety of materials derived from rocks through which the tuff forming magmas made their way to the surface, or that were picked up on the surface by the overriding ash flow have been observed by the authors in the course of studies of many hundreds of thin sections of welded tuffs from different regions of the world." They also stated that "Many tuffs show a wide range in the proportion of anorthite in the plagioclase crystals, and the more calcic ones are alien materials." They (1961, Figs. 21, 25, 43, 80, 81, 91) illustrated many examples of such xenoliths and xenocrysts in ash-flow tuffs.

Specialists other than volcanologists also are keenly aware that ash-flow tuffs such as the Toba and Bishop generally contain xenocrysts. Geochronologists, who determine the isotopic age of mineral concentrates from ash-flow tuffs such as the Bishop Tuff, have long been aware of this fact (Curtis, 1966). Naeser and others (1981) summarized the problem of contamination in ash-flow tuffs and distal volcanic ash beds and the problem contamination presents in obtaining reliable ages. For example, potassium feldspar (sanidine) concentrates from ash-flow tuffs of the Bishop Tuff yielded ages of 0.9 to 1.2 Ma (Evernden and Curtis, 1965). Dalrymple and others (1965) became aware that the Bishop Tuff contains older Sierran-age feldspar plucked from the magma chamber walls or picked up by the ash flows as they were emplaced. Therefore, they (Dalrymple and others, 1965) collected only pumice lapilli, prepared pure mineral concentrates, and refined the isotopic age of the Bishop Tuff at 0.71 Ma (average of six ages). More recently, Izett (1982) and Izett and others (1988) reported 17 new K-Ar age determinations of minerals from pristine pumice lapilli of the Bishop and redetermined its age at 0.738 ± 0.003 Ma. No features attributable to shock-metamorphism were seen in the quartz or feldspar crystals used in this study of the Bishop Tuff.

I have in my collection numerous high-quality mineral concentrates of upper Cenozoic plinian pumice deposits of the Western Interior of North America, and these are available to those interested in searching for shock-metamorphic features in pyroclastic rocks. In particular, I have many mineral concentrates of the Bishop Tuff, including one consisting of about 500 g of sanidine and smaller amounts of quartz, oligoclase, allanite, and zircon separated from a 1-m-diameter pumice boulder of the Bishop Tuff (Izett, 1982; Izett and others, 1988). Significantly, no anomalous optical features of any kind were seen in the silicate mineral concentrates of this Bishop Tuff pumice boulder—no undulatory extinction, strain shadows, Böhm lamellae, mosaicism, single sets of shock lamellae, or multiple intersecting sets of shock lamellae. This suggests the silicate minerals in the Bishop Tuff that purportedly contain rare microstructures of alleged shock origin (Carter and Officer, 1989) are xenocrysts and not primary phenocrysts.

The second choice for the origin of the xenoliths and xenocrysts in ash-flow tuffs is that they were derived from the walls and conduits of silicic magma chambers. If this idea were correct, the xenoliths and xenocrysts either should be coated with glass of the magma or concentrated in glassy pumice lapilli in ash-flow tuffs. Evidence gathered during study of volcanic ash beds derived from silicic magmas bears on this question. During the last 25 years, I have studied mineral concentrates of hundreds of volcanic ash beds, some of which are not contaminated with locally derived sediment, and air-fall pumice lapilli deposits of upper Cenozoic age in the Western Interior of North America (Izett, 1981, 1982, 1988; Izett and Wilcox, 1982). My study showed that distal volcanic ash beds do not contain volcanically derived xenocrysts or xenoliths. Instead they only contain glass-mantled phenocrysts that do not exhibit anomalous extinction or shock lamellae.

Although the idea that shock features might be found in silicate minerals of wall rocks of magma chambers and conduits should be further tested, it is difficult to understand how the great pressures (>60 kb according to Short, 1968, p. 198) and high strain rates (10^6/s according to Chao and Goresy, 1973, p. 291) required to produce multiple sets of shock lamellae could be produced in high-level silicic magma chambers.

To address the question of whether silicate minerals in wall rocks of magma chambers contain shock-metamorphic features, I collected about 100 granitic and metavolcanic xenoliths, some as large as 6.0 cm, that occur in the Bishop Tuff. The xenoliths were collected from pumice pits excavated in the air-fall pumice unit of the Bishop Tuff exposed about 10 km north of Bishop, California. Study of 44 thin sections prepared from the largest of the xenoliths revealed that they do not contain shock features, as defined by the occurrence of multiple sets of shock lamellae. Many quartz grains show normal patchy and undulatory extinction, but these features are common in such granitic and metavolcanic rocks in this setting.

Finally, it should be pointed out that specialists who have comprehensively studied gigantic Pleistocene ash-flow tuffs (and related plinian pumice deposits) have not reported finding multiple sets of planar microstructures in quartz or other silicate minerals of these rocks. Intensively studied ash-flow tuffs, to cite a few, include the Yellowstone Tuff (Christiansen and Blank, 1972), the Bishop Tuff (Hildreth, 1979), the Bandelier Tuff (Smith and Bailey, 1966), and the Toba Tuff (Chesner, 1988). Furthermore, multiple sets of planar microstructures have not been reported from Oligocene to Pliocene ash-flow tuffs of the United States (Ekren and others, 1971; Byers and others, 1976; Steven and Lipman, 1976). It seems unlikely that the many excellent petrologists who have studied such pyroclastic rocks have missed seeing such obvious, startling, and unique features as multiple intersecting sets of planar microstructures (shock lamellae) in quartz or feldspar.

AMOUNT OF SHOCK-METAMORPHOSED MINERALS IN THE K/T BOUNDARY IMPACT LAYER

The amount of quartz and other clastic mineral grains in the K/T boundary impact layer was determined gravimetrically. Weighed samples from eight different K/T boundary sites in the Raton basin and one from the Teapot Dome site in Wyoming were dissolved in dilute hydrofluoric acid using the technique described previously in this paper. Sample weight (corrected water free) and the weight percent of the residue for each of the five samples are given in Table 10. Inspection of the data shows considerable variability in weight-percent residue in samples only a few meters apart, such as at the Clear Creek North site in Colorado. The amount (weight percent) of clastic mineral grains, which varies from about 0.02 to 0.70 percent, is considerably less than the amount determined by point-counter analysis of some thin sections of the impact layer, as much as 2 percent by volume. Four factors may account for the discrepancy between the point-counter and acid-digestion results: (1) some grains may have been decanted with clay during sample preparation, (2) some small grains may have been dissolved during acid treatment, (3) it is difficult to determine precisely the amount of a trace constituent by the point-counter technique unless huge numbers of grains are counted (Chayes, 1956), and most importantly (4) extraneous rock and authigenic minerals (jarosite, gypsum, and barite) may have been inadvertently included with the impact layer when the sample was collected. Whatever the cause of the discrepancy, the apparent number of clastic mineral grains obtained by acid digestion is unquestionably less than the actual number.

The number of clastic mineral grains in the impact layer also was determined by point-counter analysis of ultrathin, doubly polished thin sections. The counting showed that the total percentage of quartzite, metaquartzite, quartz, and chert grains is variable, but at some sites the total was as much as 2 percent by volume. The proportion of these grains that contain shock-metamorphic features is difficult to ascertain using the point-

TABLE 10. CLASTIC MINERAL GRAIN CONTENT OF THE K/T BOUNDARY IMPACT LAYER FROM COLORADO, NEW MEXICO, AND WYOMING*

Location	Sample wt. (g)	Wt. clastic grains (g)	Wt. percent clastic grains
Berwind Canyon (Colo.)	468	2.78	0.59
Carmel (Colo.)	595	0.26	0.04
Clear Creek North (Colo.)	3,200	7.13	0.22
Clear Creek North (Colo.)	1,700	9.22	0.54
Clear Creek North (Colo.)	850	2.77	0.33
Clear Creek North (Colo.)	438	0.63	0.14
Clear Creek North (Colo.)	76	0.02	0.03
Madrid Railroad (Colo.)	263	0.38	0.14
Madrid East (Colo.)	259	1.82	0.70
Raton (N. Mex.)	1,507	0.62	0.04
Starkville South (Colo.)	536	1.02	0.19
North Ponil Creek (N. Mex.)	56	0.01	0.02
Teapot Dome (Wyo.)	701	0.32	0.05

*Sample weight corrected for 15 percent water (see Table 2); g, gram; five different samples from Clear Creek North were sampled; Figure 2 shows location of sites in the Raton Basin.

counter method because some grains may not be oriented so that their shock lamellae can be seen and counted. Be this as it may, it is estimated that by this method a minimum of about 20 percent of the mineral grains in the K/T boundary impact layer show definite effects of shock-metamorphism; that is, they contain multiple intersecting sets of planar lamellae.

The number of clastic mineral grains in the impact layer that contain multiple intersecting sets of shock lamellae was determined by another method. One hundred randomly chosen clastic mineral grains (>100 mesh) from the impact layer at 12 sites in western North America were mounted individually on the tip of a steel needle and studied using the spindle stage of Wilcox (1959). The mineral type and the presence or absence of shock lamellae in each grain was noted. This method is more accurate than point-counter analysis of thin sections or oil-immersion grain mounts for determining the relative abundance of shocked mineral grains because grains can be rotated and all sets of shock lamellae recorded. Quartz that displayed shock mosaicism but that did not contain shock lamellae were not counted as shocked grains. The results from the 12 sites using this method are listed in Table 9, and they show that, on average, 23.6 percent of the quartz and about 6.7 percent of the metaquartzite-quartzite grains contain multiple intersecting sets of shock lamellae. Thus, about 30 percent of the clastic mineral grains in the K/T boundary impact layer in the Western Interior of North America contain multiple sets of shock lamellae.

By using the average thickness (5 mm) and concentration of impact-derived mineral grains in the impact layer in the Raton basin and by assuming that the impact layer evenly covered a minimum area of about 12.6×10^6 km^2 in North America (2,000 km radius), I calculated that the total number of shock-metamorphosed mineral grains would be about 1.2 km^3 in this area alone.

Although the proportion of shocked mineral grains in the impact layer in the Raton basin is at least several tenths of a percent, the number of shock-metamorphosed mineral grains at K/T boundary sites outside of North America is considerably less (Izett, 1987a). At some K/T boundary sites outside of North America, shock-metamorphosed minerals are exceedingly rare. For example, I did not find a single shock-metamorphosed mineral grain in any sample I collected from three New Zealand sites: Woodside Creek, Chancet Rock, and Flaxbourne River. About 5,000 grains from each of the three sites were studied. These sites have recently been studied and discussed by Strong and others (1987, Fig. 1). Bohor and others (1985) also reported that they did not find a single definitely shocked grain in a search of 500 grains recovered from samples from Caravaca, Spain, Woodside Creek, New Zealand, and two sites in Denmark: Nye Kløv and Stevns Klint. Subsequently, a few shocked grains allegedly were found by Bohor at each of these sites (Bohor and Izett, 1986). Because the content of shocked mineral grains is so low in samples from some K/T boundary sites, such as Woodside Creek, New Zealand, extreme caution must be exercised to prevent contamination during laboratory processing of samples. This is especially true for laboratories that routinely process numerous samples that contain shocked minerals.

At European K/T boundary sites in marine rocks, shock-metamorphosed mineral grains are much more abundant than at New Zealand K/T boundary sites (Izett, 1987a). A sample of the 1.0- to 3.0-mm-thick impact layer from the Caravaca, Spain, site was collected in 1987 by the author and C. L. Pillmore because this site is exceptionally complete and well exposed. I determined the amount of shocked mineral grains by point-counter analysis of the acid insoluble residue of the impact layer in immersion oil. Only two clearly shocked quartz grains (two sets of shock lamellae) were counted in a group of 1,000 grains. Shock-metamorphosed mineral grains in the impact layer at the Petroccio and Pontedazzo, Italy, sites are about as abundant as at Caravaca, Spain. However, no clastic mineral grains containing multiple sets of shock lamellae were found in samples I collected at the Zumaya and Sopelana, Spain, and Hendaye, France, K/T boundary sites. Smit and others (1988) reported that they found shocked quartz grains as large as 0.1 mm at the Agost, Spain, site. C. L. Pillmore and I also collected samples from the K/T boundary impact layer near Agost, Spain. Numerous shocked quartz grains, that is those that contain multiple sets of shock lamellae, were identified in the insoluble residue of the impact layer from this site.

SHAPE OF SHOCK-METAMORPHOSED MINERAL GRAINS

The shape of shock-metamorphosed mineral grains was determined by optical and scanning electron microscopic (SEM) methods. The grains are angular to subangular, and some grains have a tendency to be tabular and similar in shape to feldspar cleavage flakes or crystals. A tabular shape for quartz grains is unusual because of its general lack of cleavage. Many shocked quartz grains contain and break along shock-induced planar features, most of which are parallel to $(10\bar{1}3)$, the rhombohedral planes in quartz (Englehardt and Bertsch, 1969). Therefore, many shocked quartz grains are flattened parallel to a plane about 23 degrees from the basal plane (0001).

The shapes of mineral grains in the impact layer from the Clear Creek North and Berwind Canyon sites (Fig. 2) were determined by particle analysis on an SEM. The aspect ratio (length/width) and shape (Shape = $P^2/4PA$, where P = perimeter and A = area of grain) of 805 grains from the impact layer at the Clear Creek North site averaged 1.70 and 2.28, respectively. The aspect ratio and shape of 358 grains from the impact layer at the Berwind Canyon site averaged 1.82 and 1.99, respectively.

SIZE OF SHOCK-METAMORPHOSED MINERAL GRAINS

Bohor and others (1984) reported that shock-metamorphosed quartz grains in the K/T boundary claystone at Brownie Butte, Montana, were only 0.1 mm in diameter. The apparent size of the shocked minerals was smaller than some grains measured by me (0.5 to 0.6 mm) in samples from the Raton basin (Izett and Pillmore, 1985b). This apparent difference in size was of considerable interest to me because of the possibility that grain-size distribution could be used to narrow the search for the K/T boundary impact site. In addition, the size distribution might reveal dispersal mechanisms of material away from the impact site. My study of K/T boundary samples from Montana provided by C. L. Pillmore and R. H. Tschudy showed that shock-metamorphosed grains were as large as those in the Raton basin. I found shock-metamorphosed minerals concentrated in a laminated claystone (K/T boundary impact layer) that overlies the boundary claystone. Bohor and his colleagues did not find large grains because they only studied the boundary claystone and had scraped off the impact layer that contains the concentration of shock-metamorphosed minerals (B. F. Bohor, oral communication, 1986).

I determined the size of clastic mineral grains in the impact layer, some of which contain shock lamellae, by making a sieve analysis on about 9 g of mineral grains obtained by dissolving the claystone from a large sample of the impact layer collected at the Clear Creek North site in Colorado. This analysis showed that about 30 percent of the clastic mineral grains are less than 0.1 mm, 45 percent are 0.1 to 0.2 mm, 20 percent are 0.2 to 0.3 mm, and only about 5 percent are 0.3 to 0.6 mm.

Particle-size analysis of shocked and unshocked grains using an SEM was made on mineral concentrates from the K/T boundary impact layer from the Clear Creek North and Berwind Canyon sites (Fig. 2). The mean diameter of 805 grains from the Clear Creek North site was 0.24 mm, and the mean diameter of 358 grains from the Berwind Canyon site was 0.16 mm.

To obtain another estimate of the grain-size distribution of clastic grains in the impact layer, I measured the size of individual grains in oil-immersion mounts using a petrographic microscope equipped with a calibrated micrometer ocular and the point-counter technique (Chayes, 1956). The long axis of 1,500 grains encountered during traverses of the oil-immersion mounts was measured, recorded, and used as the index of grain size. Measurements were made on grains exhibiting multiple sets of shock lamellae as well as grains that did not contain such features. Statistical analysis of the size data for shocked and unshocked grains showed that they have nearly identical means. For example, the mean size of 600 apparently unshocked grains from the Clear Creek North site in Colorado is 0.17 ± 0.06 mm, whereas the mean size of 400 shock-metamorphosed grains from the same site is 0.20 ± 0.06 mm.

The size distributions of shock-metamorphosed mineral grains in the impact layer from two sites in the Raton basin (Clear Creek North and Berwind Canyon), a newly discovered site near Teapot Dome, Wyoming (Wolfe and Izett, 1987), and a site at Brownie Butte, Montana (Bohor and others, 1984) are shown on Figure 44. In addition, the size distribution of 100 shock-metamorphosed quartz grains from the Red Deer Valley site of Lerbekmo and St. Louis (1985) is also shown on Figure 44. The mean grain size for the 100 grains was determined to be 0.26 ± 0.06 mm. The mean sizes of shocked grains from the two sites in the Raton basin are 0.20 ± 0.06 mm and 0.16 ± 0.06 mm, whereas the mean sizes of grains from Wyoming and Montana are 0.14 ± 0.04 mm and 0.15 ± 0.05 mm. The size distribution shows that grains less than about 50 μm were not encountered or measured in grain mounts. Such small grains were probably lost during sample preparation, because shock-metamorphosed grains as small as 10 μm have been seen in thin sections and in concentrates of decanted clay of the impact layer.

Although the mean size of shocked quartz grains in the impact layer in western North America ranges from 0.14 to 0.26 mm, many grains are 0.2 to 0.3 mm long. Rare large grains (0.25 to 0.64 mm) are found at all sites in western North America from New Mexico to Canada (Fig. 45). The sizes of the largest shock-metamorphosed mineral grains found by me in samples from three sites in Wyoming—the Dogie Creek site of Bohor and others (1987), the Teapot Dome site of Wolfe and Izett (1987), and the Sussex site of Nichols and others (1990)—are given on Figure 45. The largest grain found by me at the Pyramid Butte, North Dakota, site of Johnson and others (1989) was only 0.25 mm, but the K/T boundary interval at this site is not representative because it was extensively burrowed and reworked during earliest Tertiary time.

In contrast, the largest shock-metamorphosed mineral grains are only 0.19 mm long at sites outside of western North America, except for a site off the North Carolina coast. Bohor and Betterton (1989) reported the size of a shocked quartz grain as 0.30 mm in maximum dimension. The grain was from a 1-cm-thick layer at the K/T boundary in core of DSDP 603B, which is located east of Cape Hatteras and east of the lower continental rise terrace (Fig. 45, CH).

The size of shock-metamorphosed quartz grains (only those having multiple sets of shock lamellae) in the impact layer at K/T boundary sites outside of North America was established by measuring grain dimensions in oil-immersion mounts. The size of shock-metamorphosed grains was determined by the point-counter technique as described above. Some previous studies aimed at charting the size of shocked quartz grains on a global scale had reported only the size of the largest shocked grains recovered at a particular site (Bohor and Izett, 1986).

A large sample of the impact layer (see Fig. 24) from the Caravaca, Spain, site was collected in 1987. A size analysis of shocked mineral grains from this particular European site was made because the 1- to 3-mm-thick K/T boundary impact layer readily can be shaved from the enclosing sedimentary rocks. Moreover, it had been established that a reasonably large number of shocked quartz grains could be recovered from the impact layer at this site. The size distribution of shocked quartz grains recovered from the sample of the impact layer collected in 1987 is shown on Figure 44. One hundred grains were measured, and the mean grain size was 0.09 ± 0.03 mm; the size range of these grains was 0.04 to 0.19 mm. Thus, the mean grain size of shock-metamorphosed quartz grains from the Caravaca, Spain, site

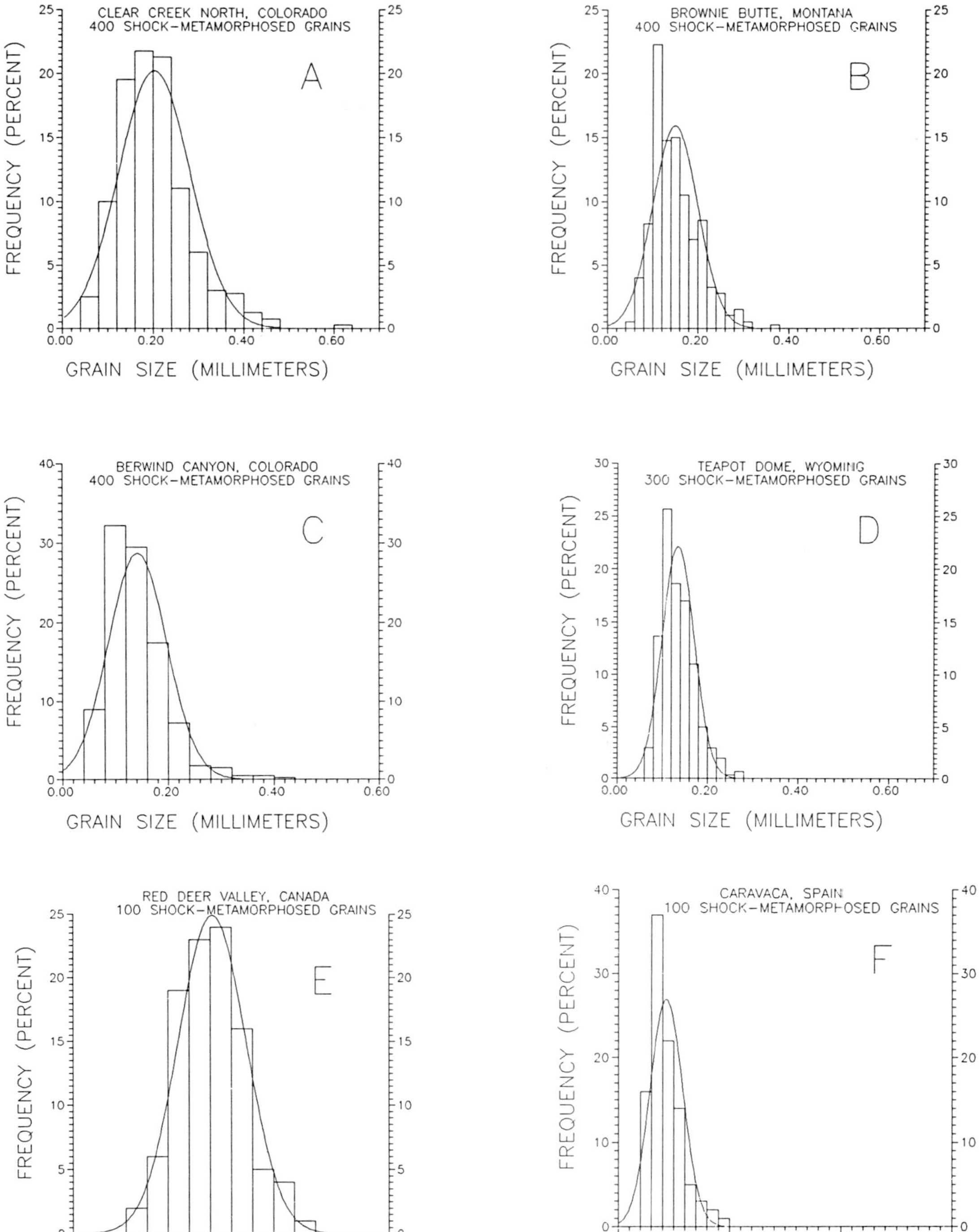

Figure 44. Frequency diagrams showing size of clastic mineral grains in the K/T boundary impact layer. Superposed chi-square distribution curves are also shown. A, 400 shocked grains from the Clear Creek North site, about 10 km south of Trinidad, Colorado. B, 400 shocked grains from the Brownie Butte, Montana, site. C, 400 shocked grains from the Berwind Canyon, about 20 km northwest of Trinidad, Colorado. D, 300 shocked grains from the Teapot Dome, Wyoming, site of Wolfe and Izett (1987). E, 100 shocked grains from the Red Deer Valley, Alberta, site of Lerbekmo and St. Louis (1985). F, 100 shocked grains from the Caravaca, Spain, site (see Izett, 1987a).

(0.09 mm) is considerably less than the mean size of shock-metamorphosed quartz grains at North American K/T boundary sites (0.14 to 0.26 mm). The largest grain measured from the Caravaca, Spain, site (0.19 mm) was about the same size as the mean grain size (0.20 mm) determined at sites in the Raton basin (see Fig. 44).

The size distribution of shocked quartz grains from sites outside of North America is as follows: Agost, Spain (mean, 0.09 ± 0.04 mm; range, 0.04 to 0.15 mm; N = 17); Stevns Klint, Denmark (mean, 0.11 ± 0.03 mm; range, 0.07 to 0.15 mm; N = 12); Petriccio, Italy (mean, 0.12 ± 0.03 mm; range, 0.07 to 0.19 mm; N = 19); GPC-3 in the north Pacific (mean, 0.12 ± 0.03 mm; range, 0.06 to 0.15 mm; N = 8). The largest shocked grain (0.19 mm) measured outside of North America was found at the Pontedazzo, Italy, site (Fig. 45). The absence of large grains could be attributed to either real differences in the original global size distribution or to the general paucity of large shocked grains. Large grains are even rare at North American sites (Fig. 44).

A size analysis of shocked mineral grains from a layer of upper Eocene impact ejecta was made to determine if their size was analogous to those in the K/T impact layer. The layer of upper Eocene impact ejecta occurs in a section of core from DSDP 612, which was drilled in the Baltimore Canyon off the New Jersey Coast. A split of the material from the core was given to me by B. F. Bohor. Thein (1987) had discovered shocked quartz grains in a 12-cm-thick layer of this coarse-grained, glassy impact material (Glass, 1987). L. W. Snee of the U.S. Geological Survey determined a $^{40}Ar/^{39}Ar$ step-heating age of 35.4 ± 0.6 Ma, reported by Obradovich and others (1989) on tektite glass fragments I hand picked from sediments of the 12-cm-thick layer in DSDP 612. The source of this layer of upper Eocene impact material is unknown, but some geologists speculated that it may have been a newly discovered early Tertiary impact crater (Jansa and Pe-Piper, 1987) in the Atlantic Ocean off Nova Scotia about 800 km north of DSDP site 612. This impact crater cannot be the source of the upper Eocene tektite glass and shocked minerals in DSDP 612 because glass from this crater formed in the early Eocene about 51 Ma (Bottomley and York, 1988).

I measured the size of 40 shocked quartz grains in the sediments of DSDP 612, and the grains range in size from 0.10 to 0.50 mm and average 0.26 ± 0.1 mm. The size distribution of shocked quartz grains in DSDP 612 is similar to the size distribution of shocked quartz grains in K/T boundary sediments in western North America.

Evidence suggests that large shocked mineral grains (0.04 to 0.19 mm) produced by the K/T impact were transported globally. I know of no other case in the geological literature documenting the global distribution of large mineral grains. It seems unlikely that such large grains could be transported by upper atmospheric winds for more than several thousand kilometers from the K/T boundary impact site. By analogy, the size of glass shards in volcanic ash beds generated during large pyroclastic eruptions generally decreases systematically downwind from their source (Izett, 1982, 1987b; Rose and Chesner, 1987). For example, the largest glass shards in the Lava Creek B volcanic ash bed are about 0.45 mm in diameter 2,000 km downwind from their source in Yellowstone National Park, Wyoming (Table 11).

Albeit there is no known common process for distributing giant particles (>50 μm) globally in the atmosphere, Betzer and others (1988) reported that they found quartz grains as large as 200 μm in dust traps in the north Pacific ocean 10,000 km downwind from their presumed source in China. A mechanism for globally transporting small-size impact material outside of the atmosphere was proposed by Jones and Kodis (1982). They concluded that particles as large as 10 μm could be transported globally above the atmosphere within a few hours after the K/T impact, but this does not explain how 100-μm or larger shocked quartz grains found worldwide were globally distributed.

LOCATION OF THE K/T BOUNDARY IMPACT

The location of the K/T boundary impact crater is controversial and not known with assurance. Emiliani and others (1981) proposed that the impact occurred in an ocean basin. If this were true, the eastern Pacific Ocean or the continental slope of western North America might be the logical place to look for

TABLE 11. MAXIMUM SIZE OF GLASS SHARDS AND CRYSTALS IN THE LAVA CREEK B VOLCANIC ASH BED (PLEISTOCENE) IN THE WESTERN UNITED STATES*

Location (County and State)	Distance from YNP (km)	Cpx	Zircon	Chev	Glass shards
Sublette Co., Wyo.	170	1.66	0.80	0.82	1.14
Fremont Co., Wyo.	200	2.45	0.82	0.71	1.13
Routt Co., Colo.	550	....	0.75	0.60	0.95
Garfield Co., Colo.	650	0.90	0.88	0.61	1.05
Jefferson Co., Colo.	700	1.00	0.68	0.55	0.86
Jefferson Co., Colo.	700	0.82	0.58	0.68	0.82
Sheridan Co., Kans.	1,020	0.32	0.27	0.32	0.76
Meade Co., Kans.	1,200	0.79	0.64	0.52	0.84
Meade Co., Kans.	1,200	0.65	0.54	0.50	0.71
Meade Co., Kans.	1,200	0.41	0.57	0.56	0.79
Meade Co., Kans.	1,200	0.48	0.43	0.50	0.60
Meade Co., Kans.	1,200	0.57	0.55	0.36	0.77
Meade Co., Kans.	1,200	0.43	0.64	0.50	0.82
Pottawattamie Co., Ia.	1,200	....	0.53	0.37	0.75
Dickens Co., Tex.	1,460	0.29	0.30	0.35	0.68
Eddy Co., N. Mex.	1,470	0.31	0.30	0.35	0.57
Johnson Co., Kans.	1,480	0.32	0.34	0.34	0.59
Presidio Co., Tex.	1,700	0.29	0.32	0.25	0.45
Bastrop Co., Tex.	2,000	0.15	0.16	0.14	0.45

*Size (in millimeters) of crystals in the ash tabulated as a function of distance from its source in Yellowstone National Park (YNP). See Izett and Wilcox (1982) for descriptions of the 0.62 Ma Lava Creek B volcanic ash bed. No data (.......); Cpx, clinopyroxene; Chev, chevkinite.

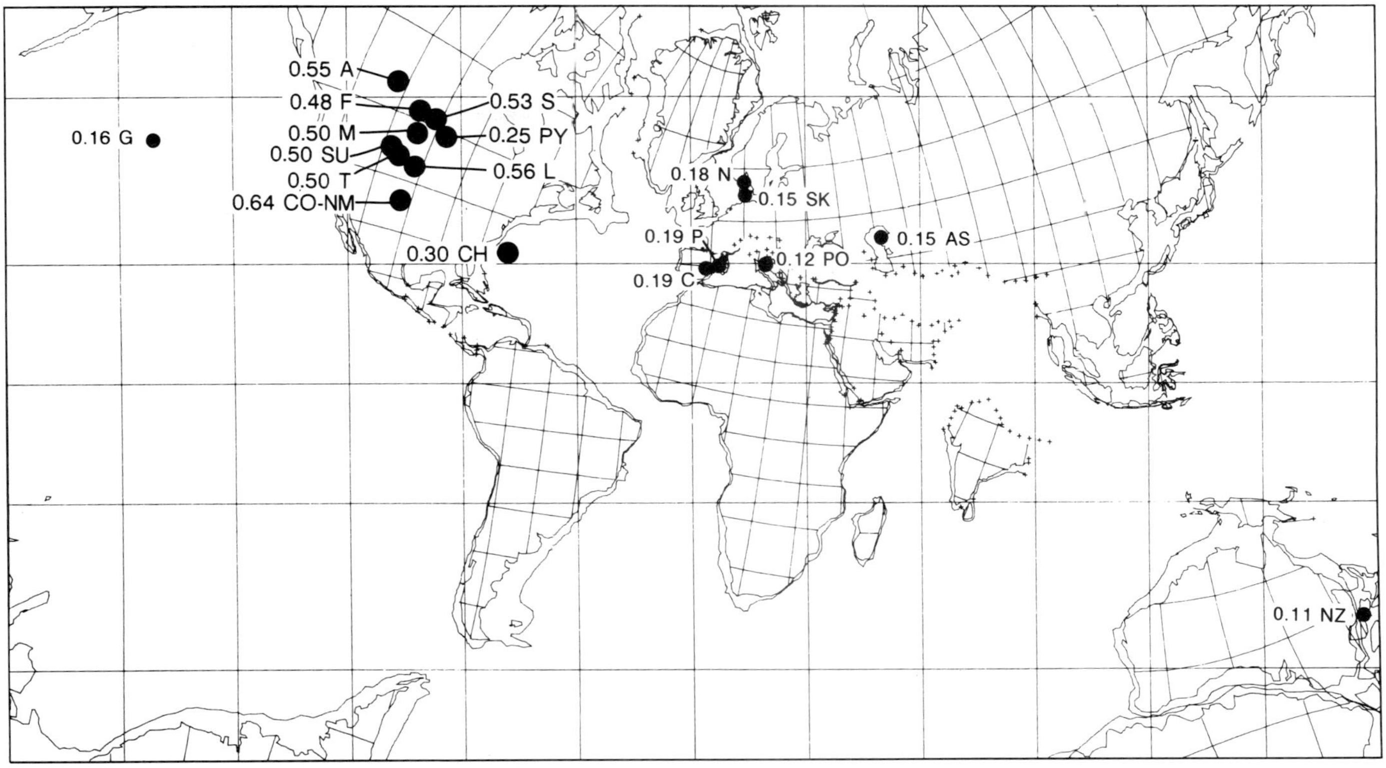

Figure 45. Paleocontinental map showing locations of K/T boundary sites on several continents and in the oceans indicating the maximum size of shocked quartz grains in K/T boundary rocks at these locations. Grain size given in millimeters. Size of locality dots indicates graphically that shocked minerals are several orders of magnitude more abundant at North American boundary sites than elsewhere in the world. Information relative to sites follows: A, Alberta, Canada (Lerbekmo and St. Louis, 1985); S, Morgan Creek site, Saskatchewan, Canada (Nichols and others, 1986); F, Frenchman River site, Saskatchewan, Canada (Lerbekmo and others, 1987); AS, Turkmenia (Badjukov and others, 1986); CO-NM, Raton basin, Colorado, and New Mexico (Fig. 2, this volume); L, Dogie Creek area, Wyoming (Bohor and others, 1987); T, Teapot Dome, Wyoming (Wolfe and Izett, 1987); SU, Sussex, Wyoming (Nichols and others, 1990); PY, Pyramid Butte, North Dakota. (Johnson and others, 1989); CH, DSDP 603B in Atlantic Ocean east of the lower continental rise terrace near Cape Hatteras off the coast of North Carolina (see Bohor and Betterton, 1989). Other sites shown by Bohor and Izett (1986) and Izett (1987a) are: M, Brownie Butte, Montana; C, Caravaca, Spain; G, GPC-3, Giant piston core, north Pacific Ocean; N, Nye Kløv, Denmark; SK, Stevns Klint, Denmark; NZ, Woodside Creek, New Zealand; P, Pontedazzo, Italy; PO, Petriccio, Italy. Base map (early Paleocene time) from Smith and Briden (1977).

the crater because of the relatively large amounts and large size of shock-metamorphosed minerals in Western Interior North American K/T boundary rocks. But the rarity and small size of shocked minerals in sediments of a giant piston core (GPC-3) in the north Pacific Ocean (Fig. 45), in relation to those in Western Interior North American boundary sites, argue against the idea that the impact occurred off the west coast of North America. The large size (0.30 mm) of a shocked quartz grain in K/T boundary sediments of DSDP 603 off the east coast of the United States (Bohor and Betterton, 1989) also suggests that the K/T impact did not occur in the Pacific Ocean.

DePaolo and others (1983) speculated that the K/T asteroid struck in an area of oceanic crust and that a substantial amount of the impact ejecta was derived from deep within the lithosphere. This conclusion was reached, in part, because the rubidium-strontium ratio ($^{87}Sr/^{86}Sr$) of the insoluble residue of claystone of the K/T impact layer at Caravaca, Spain, was markedly lower than that of marl below and above the impact layer. They assumed that the low isotopic strontium ratio of the residue of the impact layer was due to the presence of a significant proportion of air-fall impact ejecta composed of primitive oceanic lithosphere. This strontium isotopic information has been used widely to support the idea that the impact was on oceanic crust.

Unfortunately, the strontium isotopic data presented by DePaolo and others (1983) may not be used to determine the origin and provenance of the impact-layer claystone at Caravaca, Spain, for the following reason. They apparently were unaware that the impact layer at Caravaca contains tiny crystals of celestite (strontium sulfate), which have a $^{87}Sr/^{86}Sr$ ratio about equal to the value they reported for the insoluble residue of the impact layer. The celestite in the insoluble residue of the impact layer contains so much strontium with the above-cited ratio that it, in effect, swamps the strontium isotopic ratio of the rest of the insoluble residue of the impact layer (clay of terrestrial origin).

In their investigation of the K/T boundary interval rocks at Caravaca, Spain, DePaolo and others (1983) analyzed the following K/T boundary rocks: (1) acetic and hydrochloric acid insoluble residues (including sanidine spherules) of the K/T boundary impact layer and their leachates, (2) acetic and hydrochloric acid leachates of the impact layer, (3) sanidine spherules, (4) acetic and hydrochloric acid insoluble residues and leachates of marl from above and below the impact layer, and (5) Maastrichtian foraminifera from France. The ratio of strontium isotopes in authigenic celestite crystals in the K/T boundary impact layer at Caravaca, Spain, was reported by Izett (1987a) to be 0.707866 ± 0.000035 relative to a modern seawater value of 0.709182 ± 0.000006. This suggests that (1) the $^{87}Sr/^{86}Sr$ ratio for Cretaceous/Tertiary boundary seawater is about 0.707892 relative to a modern seawater value of 0.709242 (measured and reported by DePaolo and others, 1983). The $^{87}Sr/^{86}Sr$ value for the sanidine spherules is slightly more radiogenic than Cretaceous/Tertiary boundary seawater, and the $^{87}Sr/^{86}Sr$ of the insoluble residue of the impact layer is significantly lower than adjacent Cretaceous or Tertiary rocks.

Hartnady (1986) speculated that the K/T boundary impact occurred in the western Indian Ocean and formed the 300-km-diameter Amirante Basin. This idea is seemingly incompatible with mineralogic evidence found in K/T boundary rocks in western North America, including relatively large amounts and large grains of shock-metamorphosed quartz (Izett, 1987a, Fig. 10; this paper, Figs. 44 and 45).

On the basis of the large grain size (0.5 to 0.64 mm) of some shock-metamorphic minerals in the K/T boundary rocks in the Raton basin, Izett and Pillmore (1985b) suggested that the impact occurred in continental crust in North America and within a few thousand kilometers of that basin, possibly at the Manson impact structure. This suggestion was made by assuming that the decrease in size of impact material away from the K/T impact point was analogous, in part, to the decrease in size of glass shards and phenocrysts away from the sites of gigantic pyroclastic volcanic eruptions (Izett, 1987b; Rose and Chesner, 1987). Table 11 shows the size of euhedral, glass-mantled microphenocrysts and glass shards in the Lava Creek B volcanic ash bed (0.62 Ma). This ash bed was generated in Yellowstone National Park, Wyoming, during one of the largest volume pyroclastic eruptions known in the geologic record (Christiansen, 1979; Izett, 1981; Izett and Wilcox, 1982). The sizes of microphenocrysts in the ash were tabulated as a function of distance from the source vents in Yellowstone National Park, Wyoming. The largest glass-mantled microphenocrysts seen in oil-immersion mounts were measured using a calibrated micrometer ocular. Inspection of the data shows that microphenocrysts systematically decrease in size downwind from Yellowstone National Park in Wyoming. Clinopyroxene crystals are as large as 2.45 mm in Lava Creek B ash near its source, but they are only about 0.15 mm in length 2,000 km downwind near Austin, Texas.

Comparison of the data of Table 11 and Figure 45 suggests to me that the K/T boundary impact occurred in North America. The major portion of the shock-metamorphosed mineral grains produced by the impact were transported by atmospheric winds away from the impact point to their depositional sites in western North America. A small amount of small shock-metamorphosed mineral grains were transported globally by an unknown mechanism.

Prior to the suggestion of Izett and Pillmore (1985b), French (1984) had speculated that the K/T impact might have occurred on the North American continent at either the Sierra Madre or Manson impact structures in Texas and Iowa, respectively. He considered Manson more likely to be the impact site because of its larger size. French made this suggestion on the basis of the size (100 μm) of shocked minerals in K/T boundary rocks reported by Bohor and others (1984). Izett and Pillmore (1985b) showed that the size of shocked minerals was much larger than that reported by Bohor and others (1984), which strengthened the suggestion of French (1984). Although the Manson structure was originally thought to be a cryptovolcanic structure by Hale (1955, p. 35) and Hoppin and Dryden (1958, p. 699), Short and Bunch (1968, Fig. 6) demonstrated that the Manson structure clearly formed by impact processes.

Drilling in the region had established that the normal Paleozoic and Mesozoic sedimentary rock sequence is highly disturbed within an elliptically shaped area whose major axis is 40 km and minor axis is 29 km (Hale, 1955, Fig. 26; Dryden, 1955a). The exact geometry of the Manson impact structure is poorly known because of the Pleistocene till cover (~30 m), and it may be slightly larger than the limits defined by drilling records. Recent study of the Manson structure by R. R. Anderson of the Iowa Department of Natural Resources suggests that it has a diameter of 36 km (written communication, 1989).

Four published geophysical studies have been undertaken to better define the geometry of the Manson structure. Holtzman (1970) conducted a gravity survey, and he concluded that the structure is defined by a large negative gravity anomaly, but its boundary cannot be precisely located. A seismic refraction study of the Manson structure was made by Smith (1971). He suggested that the structure is circular and has a central uplift (700 m above the basement rocks) composed of crystalline rock surrounded by deformed Cretaceous shale. An aeromagnetic map of the area was prepared by Henderson and Vargo (1965). Stanley and others (1988) conducted surface and airborne geophysical studies in the Manson, Iowa, area. A magnetotelluric survey showed that the Manson structure lies along the Midcontinent Rift at the western margin of a rift-flanking linear basin. A resistivity survey clearly delineated an uplifted central block composed of high-resistivity rocks such as carbonate or granitic rock.

Several seismic reflection surveys (unpublished) have been

made across parts of the Manson structure. These data reveal the three-dimensional configuration of the structure and show that it is similar structurally to known impact structures.

According to Hartung and Anderson (1988), Paleozoic sedimentary rocks in the area of the Manson impact structure include Cambrian sandstone, limestone, and shale; Ordovician dolomite and sandstone; Devonian and Mississippian dolomite, limestone, shale, and sandstone; Pennsylvanian shale, sandstone, and coal; and Jurassic shale, sandstone, and gypsum. Mesozoic rocks include the Upper Cretaceous Dakota Sandstone and overlying Cretaceous marine shale and limestone. The total thickness of these sedimentary rocks is about 790 m (Hartung and Anderson, 1988, Fig. 2).

Igneous and metamorphic crystalline rocks near the center of the structure are uplifted about 6 km above their normal position (Hartung and Anderson, 1988). They are mainly composed of granitic and gneissic rocks that apparently form the central peak of an impact structure (Dryden, 1955a, 1955b). A geologic sketch map of the Manson, Iowa, impact structure showing the Precambrian subcrop below Pleistocene till is shown on Figure 46. This map is a modified version of a map compiled by Sims (1985) from information provided to him by R. R. Anderson of the Iowa Geological Survey.

A thick sequence (about 6 km) of Keweenawan-age Proterozoic quartzose sandstone and conglomerate associated with the Midcontinent Rift probably occurs in the subsurface in the region of the Manson structure (Hartung and Anderson, 1988). The maps of Sims (1985) and Anderson (1982) of the Midcontinent region show that quartzose clastic rocks of Proterozoic age probably extend along the Midcontinent Rift from the Lake Superior region through Iowa into Kansas.

A borehole (Manson 2A) was drilled to a depth of 146 m in 1953 in the crystalline rock near the center of the Manson structure, and a 3.8-cm-diameter core was obtained by the U.S. Geological Survey and Iowa Geological Survey. The borehole is north of Manson in the SW¼SW¼SW¼,Sec.29,T.90N.,R.31W., Pocahontas County. Records and core samples provided to me by R. R. Anderson (written communication, 1986) indicate that the depth of the borehole was 146 m. The drilling records show that about 30 m of till overlies about 58 m of highly brecciated granitic and gneissic rocks, including rare, altered, mafic-rock fragments (Dryden, 1955b). About 58 m of nonbrecciated or less brecciated granitic and gneissic rock lies below the highly brecciated interval. A second borehole (Manson 1A) was drilled in structurally disturbed shales that, according to Hoppin and Dryden (1958, p. 695), surround the granitic rock core of the structure. R. R. Anderson (written communication, 1989) believes that the Manson 1A borehole was drilled near the center of the Manson impact structure into shale downfaulted into the uplifted granitic core.

The granitic rock in the Manson 2A core was described by Dryden (1955a, 1955b), and the results of my study of the core are, in general, in agreement with his. The granitic rock is a pink to white, medium-grained rock that has a allotriomorphic texture. X-ray diffraction analysis of the granitic rock from the core shows that it is composed of quartz, potassium feldspar, and plagioclase. A small amount of kaolinite is also present. Modal analysis (1,000 points counted) of a thin section of the granite from the 144-m depth showed it chiefly consists of the following (volume percent): quartz, 34; potassium feldspar, 44; plagioclase, 16; biotite, 4; and myrmekite, 2. Accessory apatite, zircon, and magnetite also occur in the rock. Thin sections of the rock show that it contains healed fractures and has granulated grain boundaries.

The quartz is anhedral and typically occurs in grains about 0.3 mm in diameter. It has a pervasive very pale yellowish brown color in thin sections. The potassium feldspar is triclinic, as shown by x-ray diffraction analysis, and the local presence of albite-pericline twins, typical of microcline. The microcline occurs as anhedral grains as large as 4.0 mm. Its optic angle, which was estimated by the curvature of optic axis isogyres, is large ($2V_X = 80$). Dryden (1955b) mentioned that the optic angle of some potassium feldspar crystals in the granitic rock is small. As previously noted, the optic angle of some shock-metamorphosed triclinic potassium feldspar grains in the K/T boundary impact layer was lowered from near $2V_X = 80$ to $2V_X = 30$ by the shock metamorphism. The indices of refraction, which were measured in the immersion cell of a spindle stage using the focal masking technique (Wilcox, 1959, 1983), indicate that the feldspar is rich in potassium and very low in sodium ($N_X = 1.516$, $N_Y = 1.520$, and $N_Z = 1.525$). The plagioclase is calcic oligoclase as determined by measuring extinction angles (15°; X makes an angle with [010], maximum) in albite twin lamellae using a universal stage. The refractive indices of the plagioclase are $N_X = 1.534$, $N_Y = 1.539$, and $N_Z = 1.544$. The optic angle as measured on the spindle stage is large, near 90°. The biotite is generally altered to chlorite, and it occurs in smeared out masses.

The quartz and potassium feldspar (Fig. 47) in the granitic rock pervasively contain microstructures consisting of multiple intersecting sets of shock lamellae and mosaic extinction typical of shock-metamorphosed minerals (Short and Bunch, 1968, Fig. 6). The quartz contains opaque inclusions similar to those in quartz in the K/T boundary impact layer (compare Figs. 41 and 47A). The oligoclase, less commonly than the potassium feldspar, also contains shock lamellae. The potassium feldspar consistently shows mosaic extinction and poorly developed shock lamellae (Fig. 47B). Albite twin lamellae in the plagioclase are locally bent and dislocated. The oligoclase is partially altered to kaolinite and contains flecks of sericite. No thetomorphic glass was seen in either the pervasively shocked quartz or feldspars during study of five thin sections of the granitic rock of the Manson core. This observation is of interest because no thetomorphic glass has been seen in the shocked quartz or feldspars in the K/T boundary impact layer.

The gneiss in the Manson 2A core was described by Dryden (1955a, 1955b), and the results of my study of the core are in general agreement with his descriptions. I found the gneiss to consist mainly of quartz, plagioclase, and biotite. Locally it contains hornblende, pink garnet, sphene, epidote, magnetite, and

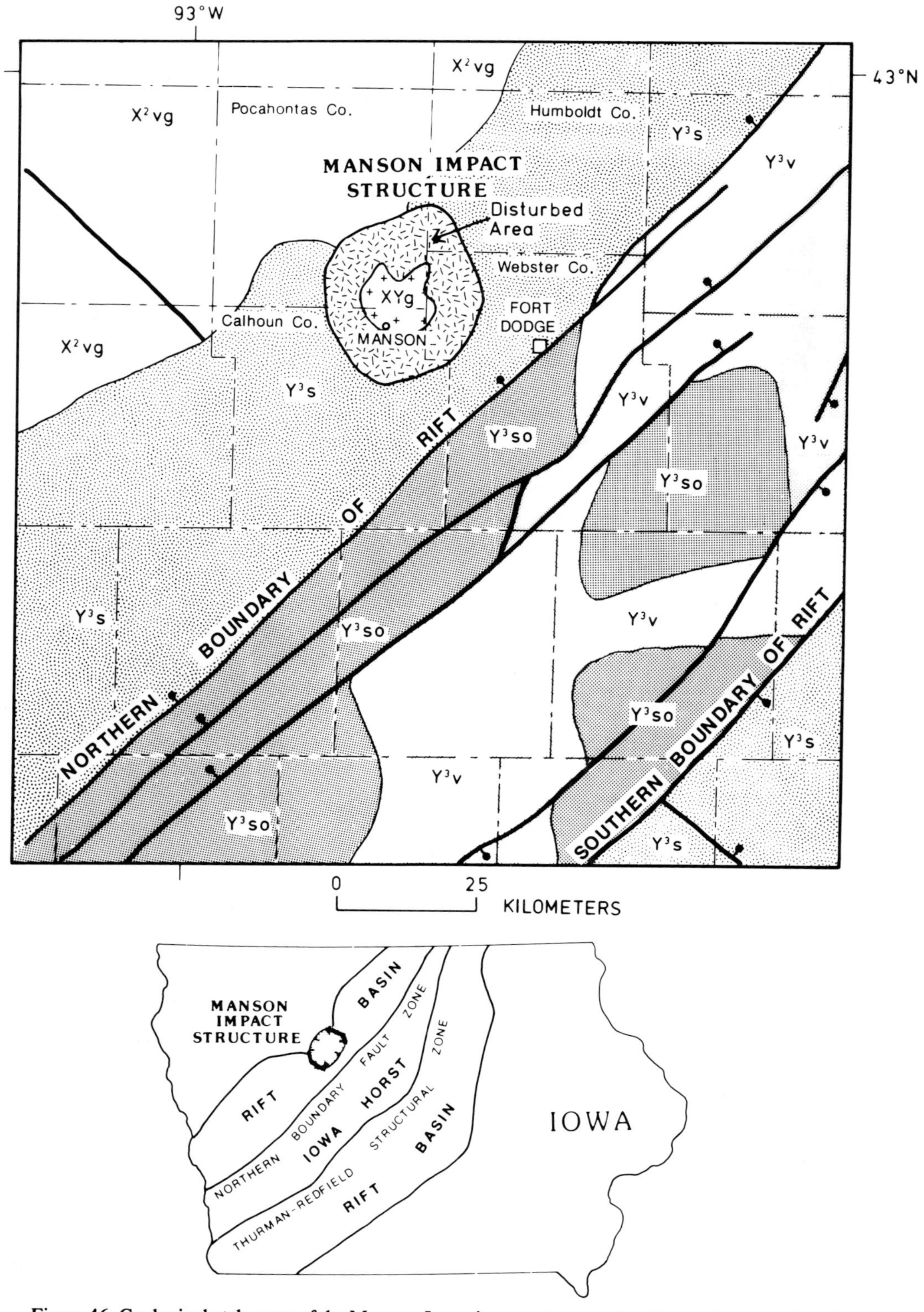

Figure 46. Geologic sketch maps of the Manson, Iowa, impact structure. Small map shows the location of the Manson impact structure in relation to the Midcontinent rift (modified from Anderson, 1982). Large map shows Precambrian basement rocks in the Manson, Iowa, region (modified from Sims, 1985). Lower Proterozoic rocks: X^2vg, granitoid and volcanic rocks undivided. Lower and middle Proterozoic rocks: XYg, granite and gneiss. Middle Proterozoic rocks: Y^3v, basalt, rhyolite, and volcanic and sedimentary rocks; Y^3so, red lithic sandstone, siltstone, and mudstone; Y^3s, red to buff quartzose sandstone, siltstone, and mudstone.

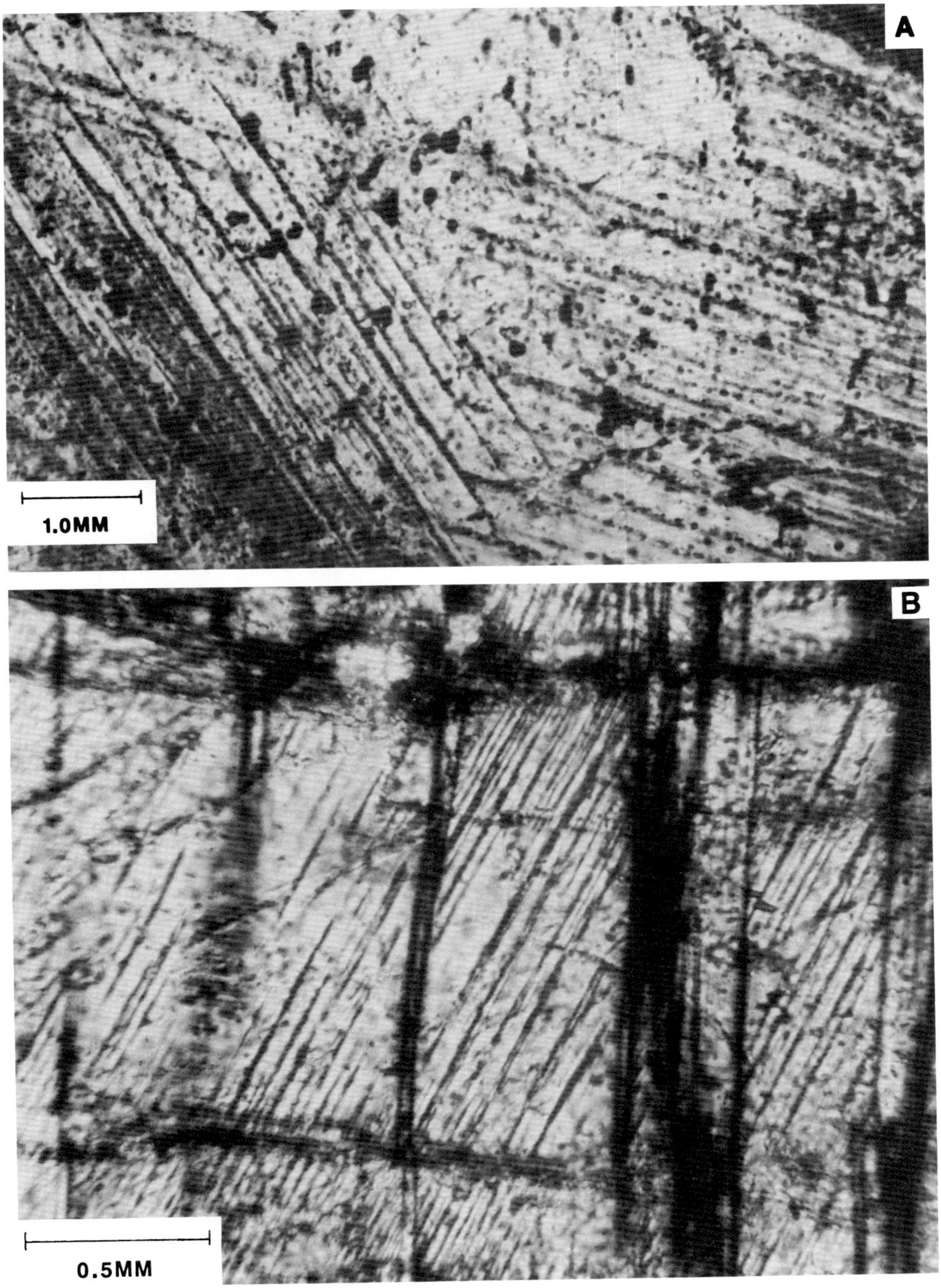

Figure 47. Photomicrograph showing shock lamellae in quartz and feldspar in granitic rock from the Manson 2A core hole. A, two sets of shock lamellae in quartz. Note opaque inclusions similar to those in shocked quartz in the K/T boundary impact layer (Fig. 41). Most quartz grains in thin sections of the granite show multiple sets of shock lamellae regardless of their orientation. B, shock lamellae in twin lamellae in microcline. Photographed with plane-polarized light.

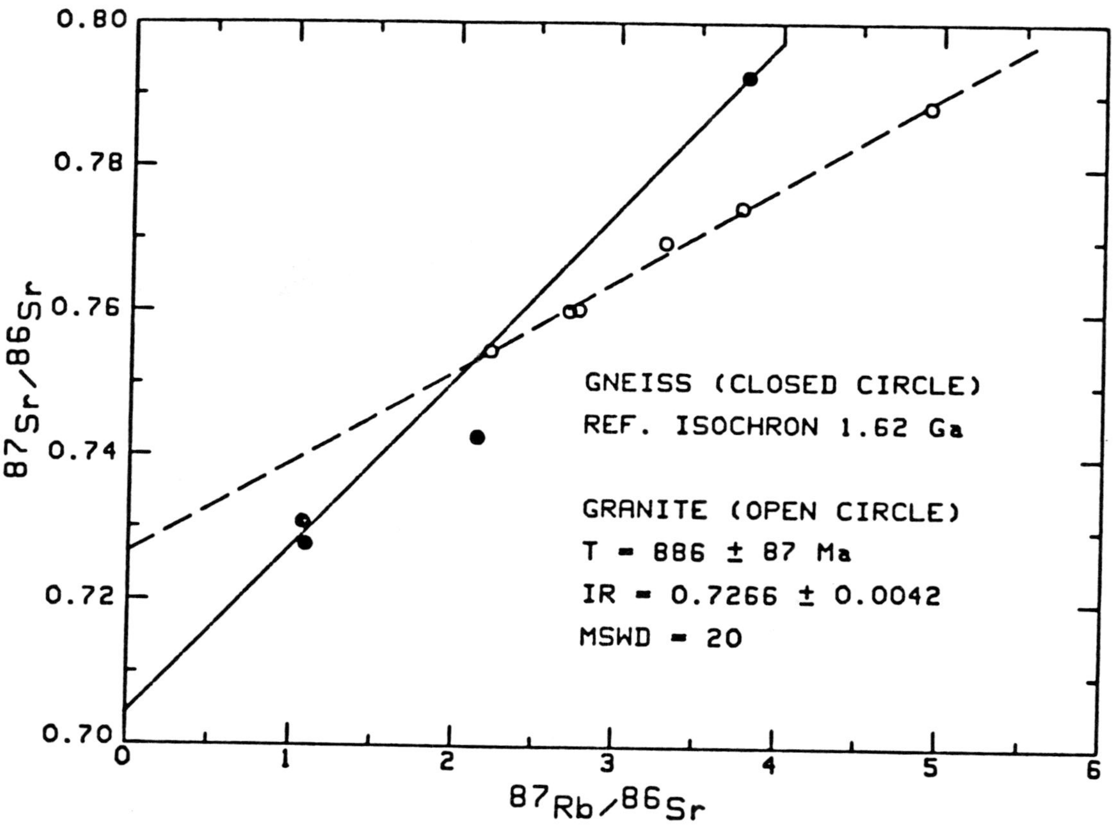

Figure 48. Rubidium-strontium isochron diagram for granitic and gneissic rocks from the Manson 2A core hole. Data from Z. E. Peterman (U.S. Geological Survey, written communication, 1987). T, time; IR, $^{87}Sr/^{86}Sr$ initial; MSWD, mean square weighted deviates.

pyrite. The quartz occurs in interlocking grains about 0.5 mm in diameter. Multiple sets of shock lamellae occur throughout the quartz. The plagioclase is generally altered to clay and its composition is difficult to ascertain; however, unaltered plagioclase was seen in two thin sections. Its composition is calcic oligoclase, as indicated by measurement of extinction angles (15°; X makes an angle with [010], maximum) in albite twin lamellae. The biotite is dark yellowish brown and relatively unaltered. An unusual mineralogical feature of the gneiss from the 138-m depth in the core is the occurrence of magnetite crystals as large as 4.0 mm in quartz-feldspar augen.

The stratigraphic age of the Manson structure is bracketed between the Late Cretaceous and Pleistocene. The Late Cretaceous age is based on the presence of the fossil clam *Inoceramus* in shale of a water well near the Manson 1A borehole (Hoppin and Dryden, 1958).

Z. E. Peterman of the U.S. Geological Survey determined the age of the granite by the rubidium-strontium, whole-rock isochron method. He analyzed six samples of granite from the Manson 2A core hole, and provided me with a plot of the data (Fig. 48). The data are aligned along an 886-Ma isochron that, according to Peterman, does not reflect the emplacement-age of the granite. Rubidium and strontium were mobilized during a metamorphic event, and the isochron was rotated to its present slope of 86 Ma (written communication, 1987). Several model ages for the granite were calculated by Peterman, using the method of Peterman and others (1985), and they average about 1,410 Ma. The average age of 1,410 Ma probably reflects the original emplacement age of the granite and is similar to the age of a granite (1,433 ± 4 Ma) from a drill core in northwest Iowa dated by Van Schmus and others (1987).

Two biotite rubidium-strontium ages of 1,070 and 1,130 Ma (old decay constants) were reported for biotite gneiss from the Manson core by Lidiak and others (1966). More recently, the isotopic age of the metamorphic rock from the Manson 2A core was determined by Z. E. Peterman. He measured the ratios of rubidium and strontium in four samples of shock-metamorphosed gneiss from the Manson 2A core hole (Fig. 48). Three of the four data points for the gneiss are aligned along a 1,620-Ma isochron. The gneiss, according to Peterman (written communication, 1987), is probably Early Proterozoic in age, near 1,800 Ma, and was overprinted during a pervasive 1,620 Ma thermal event throughout the region.

A fission-track age of a fragment of gneiss from the Manson 2A core was determined by C. W. Naeser, of the U.S. Geological Survey. I pulverized a piece of the rock from the 138-m depth and obtained an apatite mineral concentrate by heavy-liquid separation techniques. My purpose was to determine the

TABLE 12. FISSION-TRACK AGE OF APATITE CRYSTALS FROM GNEISS FROM THE MANSON 2A CORE HOLE, MANSON, IOWA*

Lab. No.	Number of grains	P_s X 10^6 t/cm^2	Fossil tracks counted	P_i X 10^6 t/cm^2	Induced tracks counted	Neutron dose X 10^{15} n/cm^2	Age (Ma)
5102	6	0.128	58	0.824	187	6.55	60.5 ± 18

Decay constant for spontaneous decay of ^{238}U (λ_f) = 7.03 X $10^{-17}yr^{-1}$.
*Analyst, C. W. Naeser; precision of age determination is ± 2 sigma; P_s, spontaneous track density; P_i, induced track density.

age of the apatite by the fission-track method and, possibly, the shock-metamorphic age of the Manson structure. As expected, the apatite was highly fractured, but no shock lamellae were seen, although the host rock is thoroughly shock-metamorphosed. The fission-track age of apatite as determined by C. W. Naeser (60.5 ± 18 Ma [± 2σ]; Table 12) was reported by Hartung and others (1986); the data are given in Table 12. Unfortunately, the analytical uncertainty of the age is large owing to the highly fractured state of the apatite. As noted previously, the apatite occurs in a Proterozoic-age gneiss, and the most reasonable interpretation for the 60-Ma fission-track age of the apatite is that it experienced a heating event probably associated with the shock metamorphism of the gneiss during the Late Cretaceous or early Tertiary.

To establish a shock-metamorphic radiometric age more precise than the fission-track method allowed, I sent M. J. Kunk and J. F. Sutter, of the U.S. Geological Survey, a sample from the 70-m depth of granitic rock from the Manson 2A corehole. Using heavy-liquid separation techniques, they separated potassium feldspar from the sample and determined its age using the $^{40}Ar/^{39}Ar$ step-heating method. The age spectrum of the feldspar was reported by Hartung and others (1986). They (p. 30) argued that the Proterozoic granitic rock containing the potassium feldspar underwent a heating event "... less than, but not much less than, 70 [m.y.] ago." Kunk and others (1987, 1989) made new $^{40}Ar/^{39}Ar$ age determinations on shock-metamorphosed potassium feldspar (microcline) from granitic rock at the 131-m depth in the Manson 2A core hole. They concluded that the microcline underwent severe argon loss about 66 Ma. The shock-metamorphic age of the microcline, and arguably the age of the Manson impact structure, is 65.7 ± 1.0 (± 2σ) Ma (Kunk and others, 1989).

In summary, several lines of evidence are compatible with the idea that the Manson, Iowa, impact structure is the place where the K/T boundary asteroid struck the Earth. The evidence includes (1) the proximity of the Manson, Iowa, impact structure and Western Interior North American K/T boundary sections that record a profound pollen break at the exact K/T boundary—apparently this pollen break was limited to the Western Interior of North America (Tschudy and Tschudy, 1986); (2) the compatibility of the composition of known and inferred subsurface rocks in the Manson area (including Proterozoic granite, gneiss, quartzite, and metaquartzite overlain by Phanerozoic sedimentary rock) and of shock-metamorphosed minerals in K/T boundary rocks; (3) the deficiency of significant amounts of thetomorphic glass in shock-metamorphosed quartz and feldspar in the impact layer and in granitic and gneissic rocks of the Manson 2A core; (4) the nearness of the Manson impact structure to western North American K/T boundary sites that contain the most abundant and largest shock-metamorphosed mineral grains in the K/T boundary impact layer; (5) the magnitude of the Manson impact structure, a minimum of about 36 km (Fig. 46); and (6) the concordance of the shock-reset isotopic age of ~66 Ma of potassium feldspar (Kunk and others, 1987, 1989) in granitic rocks of the central uplift of the Manson structure and the isotopic age of the K/T boundary (~66 Ma).

Ciskowski (1988) measured the paleomagnetic polarity of crystalline rocks (granite and gneiss) from the central uplift of the Manson impact structure and found that the alternating field demagnetization characteristics of three samples from different depths in the Manson 2A core hole have only normal polarity. This data, taken at face value, would exclude the Manson impact structure as the site where the K/T impact occurred because the K/T boundary occurs in sedimentary rocks whose polarity is reversed (Shoemaker and others, 1987). Although Ciskowski's data are interesting, much more detailed paleomagnetic study of the Manson crystalline rocks is required to establish their polarity. The normal polarity that Ciskowski measured may simply be an overprinted viscous component of magnetization acquired after the Manson impact.

K/T BOUNDARY IMPACT CRATER SIZE

Although considerable information, summarized in a preceding section of this paper, suggests that the impact structure near Manson, Iowa, may be the location of the K/T boundary impact, its diameter (~36 km) is smaller than the crater size (200 km) estimated for the K/T event by R.A.F. Grieve of the Canadian Geological Survey (in Alvarez and others, 1980, p. 1105). He estimated the size of the K/T crater on the basis of the mass and density of the K/T asteroid (10 ± 4 km) calculated by Alvarez and others (1980). However, the exact relation between the diameter of an asteroid and its impact crater is imperfectly known (Roddy and others, 1987). So-called "scaling laws" have been developed that attempt to relate asteroid and crater diameters, but these scaling laws are open to serious limitations, as pointed out by Schmidt and Holsapple (1982). Among the variables that control the diameter of a crater formed by an asteroid striking the Earth are its size, velocity, shape, density, and entry angle. In addition, the physical properties of the target material are important. It is obvious that most of these variables are either unknown or imperfectly known for the K/T asteroid, and thus, estimates that attempt to relate asteroid and impact crater diameters must be used with caution and restraint.

The size and composition of the K/T asteroid are two of the crucial parameters that must be established if the diameter of its associated impact crater is to be calculated with reasonable confidence. Alvarez and others (1980) used four different methods to calculate the size of the K/T asteroid, but two of these methods could not set close limits on the mass of the K/T asteroid.

The first method was based on the amount of iridium in K/T boundary interval rocks in Italy and Denmark. They (1980, p. 1105) assumed that the K/T asteroid was chondritic in composition and contained a mass fractional abundance of iridium equal to 0.5×10^{-6} (see Ganapathy, 1980, Table 2). By using a value of 8 ng/cm^2 for the surface density of iridium in the K/T boundary interval at Gubbio, Italy, and by using a value of 5.1×10^{18} cm^2 for the surface area of the Earth, they arrived at a calculated mass of 7.4×10^{16} g for the asteroid. They divided the calculated mass of the asteroid by the so-called "Krakatoa fraction" (0.22), which resulted in a mass of 3.4×10^{17} g for the asteroid. If the asteroid had a density of 2.2 g/cm^3, its diameter would be 6.6 km. A diameter of 14 km was calculated by using the measured surface density of iridium (80 ng/cm^2) from the Stevns Klint, Denmark, site.

The size of the asteroid was markedly increased by their use of the Krakatoa fraction. Alvarez and others (1980) drew attention to historical accounts of the 1883 eruption of the Krakatoa volcano. About 78 percent of the particulate material lofted into the atmosphere stayed in the source area; the remaining 22 percent was injected into the stratosphere and allegedly stayed aloft for 2.0 to 2.5 years. Supposedly, this material was transported globally and caused remarkable sunsets for several years following the eruption. The Alvarez team used the Krakatoa fraction, for a lack of other information, as a method for partitioning proximal and distal impact ejecta. Toon and others (1982, p. 188) noted that particulate silicate dust produced during large volcanic eruptions, such as Krakatoa, settles out within a few months following an eruption, and therefore, the use of the Krakatoa analogy is inappropriate.

If the Krakatoa fraction were not used, the calculated diameter of the K/T asteroid, using the iridium data of Alvarez and others (1980) from Gubbio, Italy, would be 4.0 km. Using the iridium data of Alvarez and others (1980) for the boundary claystone at Stevns Klint, Denmark, the calculated diameter of the K/T asteroid would be 8.8 km. If the average iridium surface density of 50 ng/cm^2 for the Raton basin were used and the mass fractional abundance of iridium in the K/T asteroid were equal to 0.5×10^{-6}, the calculated diameter of the K/T asteroid would be 7.5 km. If the mass fractional abundance of iridium in the K/T asteroid were 1.0×10^{-6} (upper limit given by Palme, 1982, Fig. 1), the calculated diameter of the asteroid would be only 5.8 km.

It is important to recognize that the above estimates of asteroid size are based on the premise that it was of chondritic composition and contained about 500 ppb iridium. However, Ganapathy (1980, p. 922) concluded that the abundance pattern of noble metals in the boundary clay does not exactly match that in CI chondrites. Similarly, Palme (1982, p. 230) observed that "The pattern of siderophiles [in the boundary claystone layers] does not match the siderophile element pattern of unfractionated meteorites (chondrites)." On the basis of ratios of osmium isotopes, Luck and Turekian (1983) concluded that the asteroid was composed of nickel-iron. Kyte and others (1985) showed that the abundance ratios of platinum-group metals for individual samples of boundary interval rocks at Caravaca, Spain, are not strictly chondritic. They noted, however, that if the samples were treated as a whole they would approach chondritic abundance ratios.

The second method used by Alvarez and others (1980) to estimate the asteroid size was based on data for earth-crossing asteroids and the impact-cratering record of the Earth. They predicted that 10-km-size asteroids should intersect the Earth, on average, about every 100 m.y. This information seems to be compatible, approximately, with the impact record of the Earth (Shoemaker and others, 1988) and the frequency of major mass-extinction events (Sepkoski, 1988).

The third method that Alvarez and others (1980) used to estimate the size of the K/T asteroid was based on the assumption that the 1-cm-thick boundary claystone at Gubbio, Italy, and Stevns Klint, Denmark, is air-fall impact material and that it occurs globally. They assumed that (1) the surface density of the boundary claystone is 2.5 g/cm^2; (2) the asteroid had a density of 2.2 g/cm^3, and the fraction of the material in the claystone of asteroidal origin is 0.5; (3) the ratio of crustal mass to asteroid mass is about 60; and (4) the Krakatoa fraction is 0.22. By using the surface area of the earth (5.1×10^{18} cm^2), they calculated that the asteroid was 7.5 km in diameter. However, their assumption that the K/T boundary claystone at Gubbio and Stevns Klint is air-fall impact material may not be justified, and accordingly, their estimate of asteroid size was incorrect. For example, Izett (1987a) demonstrated that the fallout layer at Caravaca, Spain, and Stevns Klint, Denmark, is only about 1 to 3 mm thick. Moreover, the clay minerals that compose the boundary claystone may not be impact derived. Instead, they may be normal marine clay, as suggested by Rampino (1982), Rampino and Reynolds (1983), Johnsson and Reynolds (1986), Schmitz (1988), and Elliott and others (1989).

The fourth method that Alvarez and others (1980, p. 1106) used to estimate the size of the K/T asteroid was, according to them, ". . . not able to set close limits on the mass of the incoming asteroid." This method was based on a requirement of their impact-extinction theory that the volume of impact-generated dust (and opacity of the sky) be considerably greater than that produced during the Krakatoa eruption of 1883.

Several estimates of the transient K/T crater size have been made using the size (10 km) of the asteroid as calculated by Alvarez and others (1980) as follows: 60 to 100 km by O'Keefe and Ahrens (1982, p. 104), 50 to 70 km by Melosh (1982, p. 125), and 66 km by Schmidt and Holsapple (1982, p. 93). On the basis of computer simulations, Roddy and others (1987) concluded that a 10-km-size asteroid traveling at 20 km/s would form a transient crater about 80 km in diameter in continental crust and about 105 km in diameter in oceanic crust. They also indicated that at the oceanic site the upper part of the mantle may have been melted, but none of it was ejected from the crater. These transient craters would be enlarged by secondary processes, but according to Schmidt and Holsapple (1982, p. 99), the "... degree of enlargement remains in debate." Roddy and others (1987, p. 533) calculated that "... no mantle rock was melted or ejected at the continental site." If the diameter of the K/T asteroid were only about 5 km and its velocity about 25 km/s, the transient crater would be about 40 km (Schmidt and Holsapple, 1982, Fig. 6), about the same size as the Manson impact structure.

It is important to keep in mind that the size and velocity of an asteroid are not the only parameters that have a bearing on the diameter of a crater formed by impact of the asteroid with the Earth. Just as important is the angle of incidence of the asteroid with the Earth. The above-cited estimates of the K/T boundary crater size were based on vertical incidence. Schultz and Gault (1982, p. 173) pointed out that about 50 percent of all asteroids have angles of incidence of about 45° from the horizontal. Roddy and others (1987, p. 527) stated that "... vertical impact angles require slightly lower kinetic energies than do lower impact angles to form the same size crater."

The ratio of excavated crustal material relative to asteroidal material was estimated at 60 by Alvarez and others (1980, p. 1106), 10 to 100 by O'Keefe and Ahrens (1982, p. 118), 12 by Schmidt and Holsapple (1982), and 135 by Roddy and others (1987). Melosh (1982, p. 126) concluded that the mass of material injected into the atmosphere was equal to the mass of the asteroid. Schultz and Gault (1982, p. 173) stated that "... only a small amount of the available fine-size ejecta fraction produced by such an impact may be distributed globally." Clearly, there is no consensus on the amount of target rock lofted into the atmosphere as a result of the K/T impact.

It could be argued that the Manson, Iowa, impact structure is not the K/T boundary impact site because it is too small. Many impact structures are as large or larger than the Manson structure (Grieve, 1987), but, unlike the Manson, the other large impact structures are not apparently associated with mass-extinction events. Inherent in this reasoning are the assumptions that the mass of the K/T boundary asteroid is known and that it struck with near-vertical incidence. Neither of these assumptions is warranted, as pointed out previously in this paper. If the angle of incidence of the K/T asteroid were low, a smaller crater would be formed as compared to one formed by a high strike angle. The atmosphere effects, including the generation of large volumes of nitrogen oxides (Prinn and Fegley, 1987), of a low-angle strike probably could be enormous. Thus, the killing mechanism for the K/T extinction event may not have been the lofting of large volumes of dust-size target materials into the atmosphere and the resulting global darkness as predicted by Alvarez and others (1980). Instead, the killing mechanism may have been the generation of large volumes of nitric acid produced in the atmosphere from nitrogen oxides. The resulting acid rain would have a disastrous effect on plant life, especially in the Western Interior of North America and the biota of the photozone of the oceans. Recently, O'Keefe and Ahrens (1988, 1989) speculated that the killing mechanism may have been carbon dioxide produced by the K/T asteroid impacting carbonate rocks in a shallow marine environment. The thick section of Paleozoic carbonate rocks in the Manson, Iowa, area is more appealing for their scenario because the impact probably occurred in a continental rather than an oceanic setting as discussed previously in this paper.

K-AR AGE OF THE K/T BOUNDARY

The commonly cited radiometric age of 65 Ma for the K/T boundary is based presumably on the Phanerozoic time scale of Harland and others (1964, p. 260). One of two ages used by those authors was a K-Ar age determination on plagioclase (64.8 Ma) obtained from several rounded dacite cobbles that occur 11 m above the K/T boundary in the Denver Formation at South Table Mountain east of Golden, Colorado (Evernden and others, 1964, p. 187). The boundary lies above dinosaur remains and below a mammalian fauna (Puercan) of Paleocene age (Brown, 1943). In 1970, J. D. Obradovich and I collected samples from a layer of monolithologic dacite cobbles on South Table Mountain that occurs 21 m above the Puercan mammal site described by Evernden and others (1964). The layer of cobbles is presumably 10 m stratigraphically higher than the collection made by Evernden and others (1964). A K-Ar age determination (Obradovich and Cobban, 1975) made on biotite from the dacite cobbles yielded an age of 64.3 Ma, concordant with the age of plagioclase-bearing (no biotite?) dacite cobbles dated by Evernden and others (1964). These ages recalculated using decay constants recommended by Steiger and Jäger (1977) are about 2.5 percent older than the previously calculated ages, that is, about 65.8 ± 1.4 Ma (Obradovich, 1984). Because these age determinations (~65 Ma) were made on rounded dacite cobbles, they represent a maximum age for the K/T boundary, as emphasized by Evernden and others (1964). Nevertheless, 65 Ma may also be the age of the K/T boundary if the dacite was transported and deposited shortly after its eruption.

The second radiometric age used by Harland and others (1964, item 199, p. 396) to date the K/T boundary was based on K-Ar ages of biotite and sanidine from a bentonite in the Ardley or Pembina coal zone of presumed uppermost Maastrichtian age.

The Ardley coal seam is now considered to be of early Paleocene age (see Lerbekmo and others, 1979, Fig. 1). The apparent age of this bentonite as determined by Folinsbee and others (1961) was 63 to 64 Ma based on decay constants used at that time.

Using the conventional K-Ar method, Shafiqullah and others (1964) also dated sanidine and biotite separated from bentonite beds just above the K/T boundary in Canada and Montana. The age of the K/T boundary based only on the sanidine ages from the bentonite beds and decay constants then in use was 63.4 ± 1.0 Ma. R.S.J. Lambert (in Harland and Francis, 1971, p. 79) reevaluated the geochronologic data and placed the age of the bentonites at 64.0 ± 1.0 Ma. The 64.0-Ma age for the bentonite layers just above the K/T boundary recalculated using decay constants recommended by Steiger and Jäger (1977) is about 65.6 Ma.

Lerbekmo and others (1979) reported somewhat younger K-Ar ages for sanidine from bentonites in Alberta, Canada. They dated 11 sanidine concentrates from four bentonite layers in an 18.5-m-thick interval that overlies the Nevis coal bed in the Red Deer Valley area of Alberta, Canada. This coal bed directly overlies the K/T boundary claystone and impact layer. Shock-metamorphosed quartz grains have been found in the impact layer (Izett, 1987b), and an iridium abundance anomaly was detected in the claystone and overlying the Nevis coal bed (Lerbekmo and St. Louis, 1985). They (1979) reported a mean age of 63.1 ± 0.5 (2σ) Ma (new decay constants) for the four bentonite layers. Considering all sources of error, they estimated the uncertainty of the K-Ar ages to be at least ± 3.0 Ma.

Baadsgaard and Lerbekmo (1983) dated mineral concentrates from a bentonite that occurs just above the K/T boundary in eastern Montana. They dated zircon by the U-Pb method and biotite and sanidine by the Rb-Sr method. The zircon yielded an age of 63.9 (+0.6, −0.8) Ma, and the biotite and sanidine gave an isochron age of 63.7 ± 0.3 Ma.

Obradovich (1984) and Obradovich and Sutter (1984) K-Ar dated splits of three sanidine concentrates originally K-Ar dated by Lerbekmo and others (1979) at 63.1 ± 0.5 Ma. Two of the sanidine concentrates were from a bentonite that lies 17 m above the palynological boundary and about 19 m above the K/T boundary claystone in the Red Deer Valley area of Canada, and the third sanidine concentrate was from a bentonite that is about 1 m above the K/T boundary in eastern Montana. Obradovich and Sutter used the $^{40}Ar/^{39}Ar$ step-heating method, instead of the conventional potassium-argon method used in previous studies (Lerbekmo and others, 1979), because it provides considerably more precision. Obradovich (1984) reported that the bentonite closest to the K/T boundary (1 m) dates from 66.0 ± 0.54 Ma.

Recently, Baadsgaard and others (1988) gave an average age (weighted mean) by the Rb-Sr, U-Pb, and K-Ar methods of 64.3 ± 1.2 Ma for bentonites about 1.0 m above the K/T boundary at three sites in Alberta and Saskatchewan, Canada, and Montana. Biotite and sanidine K-Ar ages were somewhat older (65.4 Ma) than the weighted mean age of 64.3 Ma (Baadsgaard and others, 1988, Table 7).

The cause for the difference in K-Ar ages for the K/T boundary as measured on a high-precision mass spectrometer in Reston, Virginia, by Obradovich and Sutter (~66 Ma) and those measured on a mass spectrometer in Alberta, Canada, by Baadsgaard and his colleagues (64 to 65 Ma) has yet to be determined, but was discussed by Obradovich and Sutter (1984).

On the basis of the foregoing discussion, I prefer to use 66 Ma for the numerical isotopic age of the K/T boundary, rather than the commonly cited 65 Ma because of the greater precision of ages determined by Obradovich (1984) and Obradovich and Sutter (1984), using the $^{40}Ar/^{39}Ar$ step-heating method.

Thus, the 66.0 ± 0.54 Ma $^{40}Ar/^{39}Ar$ age of sanidine obtained from bentonite beds at the K/T boundary (Obradovich, 1984; Obradovich and Sutter, 1984) is concordant with $^{40}Ar/^{39}Ar$ age of 65.7 ± 1.0 Ma of shock-metamorphosed microcline from the Manson, Iowa, impact structure (Kunk and others, 1987, 1989). These age determinations were made in the same laboratory using the same hornblende standard (MMhb-1) as a flux monitor. Accordingly, the Manson, Iowa, impact structure is a prime candidate for the location of the K/T impact based on the geochronologic correlation.

CONCLUSIONS

Study of K/T boundary interval rocks in the Raton basin of Colorado and New Mexico has resulted in the following conclusions:

1. Two microstratigraphic units, the K/T boundary claystone and K/T boundary impact layer, mark the K/T boundary and form a couplet at 20 sites in the Raton basin. To the north, these units have been identified and studied at several sites in Wyoming, Montana, and Canada.

2. The lower unit of the couplet, the K/T boundary claystone, is generally a 1- to 2-cm-thick kaolinitic tonstein. Where unreworked it is essentially devoid of quartz and other mineral grains. At some sites in the Raton basin the macroscopic texture of the claystone is fragmental, and at two sites in Wyoming, the unit contains centimeter-size pellets of kaolinite, indicating it is composed of reworked material. The claystone typically contains macerated plant material and locally carbonaceous structures thought to be paleoroots. The microscopic texture of the claystone is generally spherulitic where it was not reworked, and this texture probably did not originate from the direct replacement-alteration of impact or volcanic glass.

3. In the main, the trace-element chemical composition of the K/T boundary claystone differs slightly from most kaolinitic tonstein layers associated with coal beds in the Raton basin, but its composition is similar to average North American shale. The iridium content of the claystone is as much as 66 times less than that in the overlying impact layer and carbonaceous-rich rocks. The claystone contains only a few parts per million of nickel,

suggesting that the impact layer does not contain a large asteroidal component. Although the origin of the unit is uncertain and controversial, petrographic and chemical evidence suggests it is not composed of glassy altered impact material.

4. Kaolinite and goyazite spherules occur in the K/T boundary claystone, but not in the overlying K/T boundary impact layer. Iron-oxide microspherules occur in both the boundary claystone and impact layers. Stratigraphic relations and thin-section study of the kaolinite and goyazite spherules indicate that they formed by the infilling of spherical and odd-shaped cavities in the claystone. The origin of the spherical and odd-shaped cavities is unknown, although considerable evidence suggests they are not of impact origin.

5. The upper unit, the K/T boundary impact layer, consists mainly of laminated claystone that contains interlaminations of vitrinite and small barley-shaped pellets of kaolinite. At Wyoming K/T boundary sites the unit contains large kaolinite pellets, some of which contain fossil plant material. The SEM texture of the claystone is similar to I/S mixed-layer clay and is different from the pervasive microspherulitic texture of the boundary claystone. These data suggest that they have different origins.

6. The chemical composition of the impact layer claystone is similar to the underlying boundary claystone except that the former contains significantly less lithium and slightly more iron, potassium, barium, chromium, copper, vanadium, and zinc than the latter. The claystone contains only a few parts per million of nickel and cobalt and is much different in this respect than K/T boundary claystone layers at marine sites. The iridium content of K/T boundary rocks is generally highest in the impact layer, but at some sites, iridium occurs in carbonaceous mudstone and coal below the boundary claystone and above the impact layer. Seemingly, iridium was mobile during diagenesis of boundary interval sediments, and some iridium moved from the impact layer into adjacent carbonaceous-rich rocks where it was fixed.

7. The impact layer is considerably thinner than the 1-cm thickness that was visualized for the global fallout layer by Alvarez and others in 1980. At European K/T boundary sites, the impact layer is about 1 to 3 mm thick, and at Western Interior North American K/T boundary sites, it averages about 5 mm thick. Because the impact layer is not composed mainly of altered impact material, its thickness does not reflect its distance from the K/T impact site.

8. The contact between the K/T boundary claystone and K/T boundary impact layer separates two distinctly different lithologies—relatively homogeneous claystone from overlying laminated claystone. Locally the contact may be a diastem. Furthermore, the evidence seems to indicate that the top of the boundary claystone locally was a paleosurface upon which plants grew and the impact layer was deposited. The amount of time represented by the paleosurface or diastem is probably brief.

9. The impact layer contains a clastic mineral grain assemblage (as much as 2 percent) chiefly composed of quartz, oligoclase, and microcline and lithic fragments of quartzite, metaquartzite, chert, and granite-like mixtures of quartz, oligoclase and microcline.

10. Most impact layer mineral concentrates contain a miniscule amount of small ($<100\ \mu m$), rounded heavy minerals including zircon, allanite, tourmaline, epidote, garnet, chromium spinel, ilmenite, and metamorphic blue-green hornblende. Similar rounded heavy minerals occur in sedimentary rocks stratigraphically below and above the K/T boundary impact layer. The presence of this detrital assemblage of heavy minerals in the impact layer at continental K/T boundary sites implies that it is not composed of impact material.

11. About 30 percent of the clastic mineral grains (quartz, quartzite, and metaquartzite; not chert and chalcedony) in the impact layer in the Raton basin contain multiple sets of planar microstructures of shock-metamorphic origin. These microstructures provide clear and convincing evidence that an steroid struck Earth at K/T boundary time.

12. The lack of significant amounts of stishovite, thetomorphic glass, and significant lowering of the indices of refraction of shocked quartz grains in the impact layer suggest that quartz and other clastic minerals experienced only low shock levels (~100 kb) at the impact site.

13. Shock-metamorphosed mineral grains and an unknown phase that carried the iridium are the only impact-related components in the claystone. The impact-produced components probably were transported atmospherically away from the impact site and fell as a light sprinkling throughout central North America soon after the K/T boundary impact.

14. Shock-metamorphosed minerals are concentrated in the K/T boundary impact layer and not in the underlying K/T boundary claystone. Only one such concentration of shock-metamorphosed minerals has been found near the K/T boundary, supporting the idea that there was only one impact event.

15. Unshocked lithic grains of quartzite, metaquartzite, chert, and chalcedony are the main components of the impact layer, but they are less abundant in rocks below the boundary claystone in the Raton basin. This change takes place at the same stratigraphic position at each site in the Raton basin and at some sites in Wyoming. This evidence suggests that a considerable amount of the unshocked mineral grains may be impact target materials.

16. The impact layer contains larger shocked quartz grains (mean = 0.14 to 0.20 mm) at North American K/T boundary sites relative to boundary sites elsewhere in the world (mean = 0.09 mm). Perhaps of more importance, shocked mineral grains are much more abundant at western North America K/T sites as compared to elsewhere in the world. The compositional types of shock-metamorphosed materials in the impact layer are thought to reflect the composition of target rocks at the impact site. The fact that the clastic mineral grain assemblage in the impact layer contains a large amount of quartz, quartzite, metaquartzite, and chert indicates that the K/T impact target rocks were probably quartzose sedimentary, metasedimentary, and metamorphic

rocks. Small amounts of shocked feldspar and granite-like mixtures of feldspar and quartz suggest that the target rock was, in part, granitic.

17. The nature of the shock-metamorphosed minerals in K/T boundary rocks indicates that they are not of volcanic origin. This conclusion was reached from two different areas of research, one from the study of shock-metamorphosed minerals in K/T boundary rocks and a second from study of silicic pyroclastic volcanic rocks. A large percentage of the minerals in the impact layer are quartzite and metaquartzite whose source was sedimentary, metasedimentary, and metamorphic rocks and certainly could not have been silicic pyroclastic rocks.

18. The amount of fine-grained, atmospherically transported K/T boundary impact ejecta is considerably smaller than that estimated by Alvarez and others (1980) because the clay minerals that compose the boundary claystone are not altered impact ejecta. Therefore, the diameters of the K/T asteroid and resulting impact crater were also much smaller than the previous estimate (10 and 200 km, respectively).

19. Sedimentary rocks that contain the K/T boundary interval consist of claystone and coal deposited in a low-energy depositional regime. The lack of conglomeratic channel sandstone beds or coarse-grained sediments directly above the K/T boundary impact bed suggests that there were no major high-energy depositional events triggered by wildfires or acid rain fallout associated with the K/T impact in the Western Interior of North America.

20. The Manson, Iowa, impact structure is probably where the K/T boundary impact occurred because of (1) the compositional similarity of known and inferred Manson subsurface rocks and shocked K/T boundary minerals, (2) the large size (~36 km diameter) of the impact structure, (3) the essentially identical $^{40}Ar/^{39}Ar$ ages (~66 Ma) of shock-metamorphosed microcline from Manson granitic rocks and sanidine from bentonites that overlie the K/T boundary coal bed in Canada and Montana, and (4) the proximity of the Manson impact structure to Western Interior North American K/T boundary sites that contain relatively abundant and large grain-size minerals.

REFERENCES CITED

Adams, P. B., 1984, Glass corrosion, *in* Pye, L. D., and others, eds., Natural glasses: Amsterdam, North-Holland Physics Publishing Co., p. 193–205.

Ahrens, T. J., and Rosenberg, J. T., 1968, Shock metamorphism; Experiments on quartz and feldspar, *in* French, B. M., and Short, N. M., eds., Shock metamorphism of natural materials: Baltimore, Maryland, Mono Book Corporation, p. 59–81.

Alekseev, A. S., Barsukova, L. D., Kolesov, G. M., Nazarov, M. A., and Amanniyazov, K. N., 1986, Cretaceous/Tertiary event; Iridium distribution in Turkmenia sections, *in* Abstracts of papers, Lunar and Planetary Science Conference 17: Houston, Texas, Lunar and Planetary Institute, p. 9–10.

Alexopoulos, J., Grieve, R.A.F., and Robertson, R. B., 1987, Microscopic lamellar features in quartz from different geologic environments, *in* Abstracts of papers, Lunar and Planetary Science Conference 17: Houston, Texas, Lunar and Planetary Institute, p. 19–20.

——, 1988, Microscopic lamellar features in quartz; Discriminative characteristics of shock-generated varieties: Geology, v. 16, p. 796–799.

——, 1989, Comment and Reply: Geology, v. 15, p. 90–92.

Allen, V. T., 1932, Ordovician altered volcanic material in Iowa, Wisconsin, and Missouri: Journal of Geology, v. 40, p. 259–269.

——, 1946, Sedimentary and volcanic processes in the formation of high alumina clay: Economic Geology, v. 41, p. 124–138.

——, 1952, Petrographic relations in some typical bauxite and diaspore deposits: Geological Society of America Bulletin, v. 63, p. 649–688.

Allen, V. T., and Nichols, R. L., 1945, Clay-pellet conglomerates at Hobart Butte, Lane County, Oregon: Journal of Sedimentary Petrology, v. 15, p. 25–33.

Alvarez, L. W., 1983, Experimental evidence that an asteroid impact led to the extinction of many species 65 million years ago: National Academy of Sciences Proceedings, v. 80, p. 627–642.

Alvarez, L. W., Alvarez, W., Asaro, F., and Michel, H. V., 1979a, Extraterrestrial cause for the Cretaceous–Tertiary extinction; Experiment and theory: Berkeley, University of California, Berkeley Lawrence Report LBL–9666, 86 p.

——, 1979b, Experimental evidence in support of an extraterrestrial trigger for the Cretaceous–Tertiary extinctions: EOS Transactions of the American Geophysical Union, v. 60, no. 42, p. 734.

——, 1980, Extraterrestrial cause for the Cretaceous–Tertiary extinction: Science, v. 208, p. 1095–1108.

Alvarez, W., 1986, Toward a theory of impact crises: EOS Transactions of the America Geophysical Union, v. 67, p. 649–658.

Alvarez, W., and Montanari, A., 1985, The sedimentary deposits of major impact events: Geological Society of America Abstracts with Programs, v. 17, p. 512–513.

Alvarez, W., Alvarez, L. W., Asaro, F., and Michel, H. V., 1982, Current status of the impact theory for the terminal Cretaceous extinction, *in* Silver, L. T., and Schultz, P. H., eds., Geological implications of impacts of large asteroids and comets on the Earth: Geological Society of America Special Paper 190, p. 305–315.

Anderson, A. T., 1976, Magma mixing; Petrological processes and volcanological tool: Journal of Volcanology and Geothermal Research, v. 1, p. 3–33.

Anderson, A. T., Newman, S., Williams, S. N., Druitt, T. H., Skirius, C., and Stolper, E., 1989, H_2O, CO_2, Cl, and gas in plinian and ash-flow Bishop [Tuff] rhyolite: Geology, v. 17, p. 221–225.

Anderson, C. A., and Creasey, S. C., 1958, Geology and ore deposits of the Jerome area, Yavapai County Arizona: U.S. Geological Survey Professional Paper 308, 185 p.

Anderson, R. R., 1982, Iowa's distant past: Iowa Geology, no. 7, p. 22–24.

Archibald, J. D., and Clemens, W. A., 1982, Late Cretaceous extinctions: American Scientist, v. 70, p. 377–385.

Baadsgaard, H., and Lerbekmo, J. F., 1983, Rb-Sr and U-Pb dating of bentonites: Canadian Journal of Earth Sciences, v. 20, p. 1282–1290.

Baadsgaard, H., Lerbekmo, J. F., and McDougall, I., 1988, A radiometric age for the Cretaceous-Tertiary boundary based upon K-Ar, Rb-Sr, and U-Pb ages of bentonites from Alberta, Saskatchewan, and Montana: Canadian Journal of Earth Sciences, v. 25, p. 1088–11097.

Badjukov, D. D., Nazarov, M. A., and Suponeva, I. V., 1986, Shocked quartz grains from K/T boundary sediments, *in* Abstracts of papers, Lunar and Planetary Science Conference 17: Houston, Texas, Lunar and Planetary Institute, p. 18–19.

Baksi, A. J., 1987, Critical evaluation of the age of the Deccan Traps, India; Implications for flood-basalt volcanism and faunal extinctions: Geology, v. 15, p. 147–150.

——, 1989, Comment and Reply: Geology, v. 17, p. 89–90.

Barker, F., and Friedman, I., 1974, Precambrian metavolcanic rocks of the Tusas Mountains, New Mexico; Major elements and oxygen isotopes, *in* Siemers, C. S., ed., Silver anniversary guidebook, 25th field conference: New Mexico Geological Society, p. 115–117.

Barker, J. L., Jr., and Anders, E., 1968, Accretion rate of cosmic matter from iridium and osmium contents of deep sea sediments: Geochimica et Cosmochimica Acta, v. 32, p. 627–645.

Berggren, W. A., and Van Couvering, J. A., eds., 1984, Catastrophes and Earth history: Princeton, New Jersey, Princeton University Press, 464 p.

Berner, R. A., and Landis, G. P., 1987, Chemical analysis of gaseous bubble inclusions in amber: Geological Society of America Abstracts with Programs, v. 19, p. 587.

Besse, J., Courtillot, V., and Vandamme, D., 1989, Comment and Reply: Geology: v. 17, p. 88–89.

Betzer, P. R., and 11 others, 1988, Long-range transport of giant mineral aerosol particles: Nature, v. 336, p. 569–571.

Birkelund, T., and Bromley, R. G., eds., 1979, Proceedings Part 2, Cretaceous-Tertiary boundary events: Copenhagen, Denmark, University of Copenhagen, 262 p.

Bohor, B. F., 1977, Tonsteins as correlation tools and environmental indicators of Rocky Mountain coals: Geological Society of America Abstracts with Programs, v. 9, p. 710–711.

——, 1988a, K-T boundary claystone is a distal ejecta blanket: Fify-first annual meeting of the Meteoritical Society, p. N-8.

——, 1988b, Shocked quartz and more; Impact signatures in K-T boundary clays and claystones, *in* Global catastrophes in Earth history; An interdisciplinary conference on impacts, volcanism, and mass mortality; Abstracts Presented to the Topical Conference: Houston, Texas, Lunar and Planetary Institute Contribution 673, p. 17–18.

Bohor, B. F., and Betterton, W. J., 1989, Glauconite spherules and shocked quartz at the K-T boundary in DSDP 603 B, *in* Abstracts of papers, Lunar and Planetary Science Conference 20: Houston, Texas, Lunar and Planetary Institute, p. 92–93.

Bohor, B. F., and Izett, G. A., 1986, Worldwide size distribution of shocked quartz at the K/T boundary; Evidence for a North American impact site, *in* Abstracts of papers, Lunar and Planetary Science Conference 17: Houston, Texas, Lunar and Planetary Institute, p. 68–69.

Bohor, B. F., and Pillmore, C. L., 1976, Tonstein occurrences in the Raton coal field, Colfax County, New Mexico; New Mexico Geological Society Guidebook, 27th Field Conference, Vermejo Park, 1976: New Mexico Geological Society, p. 177–183.

Bohor, B. F., and Triplehorn, D. M., 1987, Flyash; An analog for spherules in K-T boundary clays, *in* Abstracts of papers, Lunar and Planetary Science Conference 18: Houston, Texas, Lunar and Planetary Institute, p. 103–104.

Bohor, B. F., Foord, E. E., Modreski, P. J., and Triplehorn, D. M., 1984, Mineralogical evidence for an impact event at the Cretaceous–Tertiary boundary: Science, v. 224, p. 867–869.

Bohor, B. F., Modreski, P. J., and Foord, E. E., 1985, A search for shock-metamorphosed quartz at the K-T boundary, *in* Abstracts of papers, Lunar and Planetary Science Conference 16: Houston, Texas, Lunar and Planetary Institute, p. 79–80.

Bohor, B. F., Foord, E. E., and Ganapathy, R., 1986, Magnesioferrite from the

Cretaceous–Tertiary boundary, Caravaca, Spain: Earth and Planetary Science Letters, v. 81, p. 57–66.

Bohor, B. F., Triplehorn, D. M., Nichols, D. J., and Millard, H. T., Jr., 1987, Dinosaurs, spherules, and the "magic" layer; A new K-T boundary clay site in Wyoming: Geology, v. 15, p. 896–899.

Bonnie, C. A., and Honess, A. P., 1929, Bentonite in Pennsylvania: Pennsylvania Academy of Science Proceedings, v. 3, p. 1–8.

Bottomley, R., and York, D., 1988, Age measurement of the submarine Montagnais impact crater: Geophysical Research Letters, v. 15, p. 1409–1412.

Bowles, J. F., 1986, The development of platinum-group minerals in laterites: Economic Geology, v. 81, p. 1278–1285.

Boynton, W. V., 1984, Cosmochemistry of the rare earth elements; Meteorite studies, *in* Henderson, P., ed., Rare earth element geochemistry: Amsterdam, Elsevier, p. 63–114.

Bradley, W. H., 1964, Geology of the Green River Formation and associated Eocene rocks in southwestern Wyoming and adjacent parts of Colorado and Utah: U.S. Geological Survey Professional Paper 496-A, 86 p.

Brooks, R. R., and 7 others, 1984, Elemental anomalies at the Cretaceous-Tertiary boundary, Woodside Creek, New Zealand: Science, v. 226, p. 539–542.

Brown, R. W., 1943, Cretaceous-Tertiary boundary in the Denver Basin, Colorado: Geological Society of America Bulletin, v. 54, p. 65–86.

Brownlee, D. E., 1981, Extraterrestrial components, *in* Emiliani, C., ed., The sea: New York, John Wiley and Sons, p. 733–762.

Bunch, T. E., 1968, Some characteristics of selected minerals from craters, *in* French, B. M., and Short, N. M., eds., Shock metamorphism of natural materials: Baltimore, Maryland, Mono Book Corporation, p. 413–441

Burger, K., 1985, Petrography and chemistry of tonsteins of the coal basins of western Europe and North America, *in* Cross, A. T., ed., Economic geology; Coal, oil, and gas; International Congress of the Stratigraphy and Geology of the Carboniferous, 1979: Compte Rendu, v. 9, no. 4, p. 449–466.

Byers, F. M., Carr, W. J., Orkild, P. P., Quinlivian, W. D., and Sargent, K. A., 1976, Volcanic suites and related cauldrons of Timber Mountain–Oasis Valley caldera complex, southern Nevada: U.S. Geological Survey Professional Paper 919, 70 p.

Carter, N. L., 1965, Basal quartz deformation lamellae; A criterion for recognition of impactites: American Journal of Science, v. 263, p. 786–806.

—— , 1968, Dynamic deformation of quartz, *in* French, B. M., and Short, N. M., eds., Shock metamorphism of natural materials: Baltimore, Maryland, Mono Book Corporation, p. 453–474.

—— , 1971, Static deformation of silica and silicates: Journal of Geophysical Research, v. 76, p. 5514–5540.

Carter, N. L., and Friedman, M., 1965, Dynamic analysis of deformed quartz and calcite from the Dry Creek Ridge anticline, Montana: American Journal of Science, v. 263, p. 747–785.

Carter, N. L., and Officer, C. B., 1986, Comment and Reply: Geology, v. 15, p. 90–92.

—— , 1989, Comment and Reply: Geology, v. 17, p. 477–480.

Carter, N. L., Christie, J. M., and Griggs, D. T., 1964, Experimental deformation and recrystallization of quartz: Journal of Geology, v. 72, p. 687–733.

Carter, N. L., Officer, C. B., Chesner, C. A., and Rose, W. I., 1986, Dynamic deformation of volcanic ejecta from the Toba caldera; Possible relevance to the Cretaceous-Tertiary boundary phenomena: Geology, v. 14, p. 380–383.

Chao, E.C.T., 1967, Shock effects in certain rock-forming minerals: Science, v. 156, p. 192–202.

—— , 1968, Pressure and temperature histories of impact metamorphosed rocks based on petrographic observations, *in* French, B. M., and Short, N. M., eds., Shock metamorphism of natural materials: Baltimore, Maryland, Mono Book Corporation, p. 135–158.

—— , 1976, The application of quantitative interference microscopy to mineralogic and petrologic investigations: American Mineralogist, v. 61, p. 212–228.

Chao, E.C.T., and Goresy, A. el, 1973, Shock attenuation and the implantation of Fe-Cr-Ni veinlets in the compressed zone of the 1973 Ries research drill core: Geologica Bavarica, v. 75, p. 289–304.

Chayes, F., 1956, Petrographic modal analysis: New York, John Wiley, 113 p.

Chesner, C. A., 1988, The Toba tuffs and caldera complex, Sumatra, Indonesia; Insights into magma bodies and eruptions [Ph.D. thesis]: Houghton, Michigan Technological University, 428 p.

Christiansen, R. L., 1979, Cooling units and composite sheets in relation to caldera structure, *in* Chapin, C. E., and Elston, W. E., eds., Ash-flow tuffs: Geological Society of America Special Paper 180, p. 29–42.

Christensen, L., Fregerslev, S., Simonsen, A., and Theide, J., 1973, Sedimentology and depositional environment of lower Danian fish clay from Stevns Klint, Denmark: Geological Society of Denmark Bulletin, v. 22, p. 193–212.

Christiansen, R. L., and Blank, H. R., 1972, Volcanic stratigraphy of the Quaternary rhyolite plateau in Yellowstone National Park: U.S. Geological Survy Professional Paper 729-B, 18 p.

Ciskowski, S. M., 1988, Paleomagnetism of Manson structure cores inconsistent with a K/T link, *in* Abstracts of papers, Lunar and Planetary Science Conference 19: Houston, Texas, Lunar and Planetary Institute, p. 188–189.

Courtillot, V., and 5 others, 1986, Deccan flood basalts at the Cretaceous/Tertiary boundary?: Earth and Planetary Science Letters, v. 80, p. 361–374.

Courtillot, V., and 5 others, 1988, Deccan flood basalts and the Cretaceous/Tertiary boundary: Nature, v. 333, p. 843–846.

Cullers, R. L., Chauduri, S., Arnold, B., Lei, M., Wolf, C. W., Jr., 1975, Rare-earth distributions in clay minerals and in clay-sized fraction of the Lower Permian Havensville and Eskridge Shales of Kansas and Oklahoma: Geochimica et Cosmochimica Acta, v. 39, p. 1691–1703.

Curtis, G. H., 1966, The problem of contamination in obtaining accurate dates of young geologic rocks, *in* Schaeffer, O. A., and Zähringer, J., eds., Potassium argon dating: New York, Springer-Verlag, p. 151–162.

Dalrymple, G. B., Cox, A., and Doell, R. R., 1965, Potassium-argon age and paleomagnetism of the Bishop Tuff, California: Geological Society of America Bulletin, v. 76, p. 665–674.

DePaolo, D. J., Kyte, F. T., Marshall, B. D., O'Neil, J. R., and Smit, J., 1983, Rb-Sr, Sm-Nd, K-Ca, O, and H isotopic study of Cretaceous-Tertiary boundary sediments, Caravaca, Spain; Evidence for an oceanic impact site: Earth and Planetary Science Letters, v. 64, p. 356–373.

Desborough, G. A., Pitman, J. K., and Donnell, J. R., 1973, Microprobe analysis of biotites; A method of correlating tuff beds in the Green River Formation, Colorado and Utah: U.S. Geological Survey Journal of Research, v. 1, p. 39–44.

Dryden, J. E., 1955a, A near-surface crystalline mass at Manson, Iowa, *in* Institute on Lake Superior geology: Minneapolis, University of Minnesota Center Continuation Study, unpaged.

—— , 1955b, A study of a well core from crystalline rocks near Manson, Iowa [M.A. thesis]: Ames, Iowa State University, 89 p.

Duncan, R. A., and Pyle, D. G., 1988, Rapid eruption of the Deccan flood basalts at the Cretaceous/Tertiary boundary: Nature, v. 333, p. 841–843.

Ekren, E. B., Anderson, R. E., Rogers, C. L., and Noble, D. C., 1971, Geology of northern Nellis Air Force Base Bombing and Gunnery Range, Nye County, Nevada: U.S. Geological Survey Professional Paper 651, 91 p.

Elliott, W. C., Aronson, J. L., Millard, H. T., Jr., and Gierlowski-Kordesch, E., 1989, The origin of the clay minerals at the Cretaceous/Tertiary boundary in Denmark: Geological Society of America Bulletin, v. 101, p. 702–710.

Emiliani, C., Kraus, E. B., and Shoemaker, E. M., 1981, Sudden death at the end of the Mesozoic: Earth and Planetary Science Letters, v. 55, p. 317–334.

Englehardt, W. von, and Bertsch, W., 1969, Shock induced planar deformation structures in quartz from Ries Crater, Germany: Contributions to Mineralogy and Petrology, v. 20, p. 203–234.

Englehardt, W. von, Horz, F., Stöffler, D., and Bertsch, W., 1968, Observations on quartz deformation in the breccias of West Clearwater Lake, Canada, and the Ries basin, Germany, *in* French, B. M., and Short, N. M., eds., Shock metamorphism of natural materials: Baltimore, Maryland, Mono Book Corporation, p. 475–482.

Evernden, J. F., and Curtis, G. H., 1965, The present status of potassium-argon dating of Tertiary and Quaternary rocks, *in* International Association Quaternary Research, 6th Congress, Warsaw, 1961: Lodz, Poland, Panstwowe Wydawnictwo Naukowe Oddizal w Lodzi, v. 1, p. 643–651.

Evernden, J. F., Savage, D. E., Curtis, G. H., and James, G. T., 1964, Potassium-argon dates and the Cenozoic mammalian chronology of North America: American Journal of Science, v. 262, p. 145–198.

Fahey, J. J., 1964, Recovery of stishovite from Coconino Sandstone of Meteor Crater, Arizona: American Mineralogist, v. 49, p. 1643–1647.

Fastovsky, D. E., and McSweeney, K., 1987, Paleosols spanning the Cretaceous-Paleocene transition, eastern Montana and western North Dakota: Geological Society of America Bulletin, v. 99, p. 66–77.

Fastovsky, D. E., McSweeney, K., and Norton, L. D., 1989, Pedogenic development at the Cretaceous-Tertiary boundary, Garfield County, Montana: Journal of Sedimentary Petrology, v. 59, p. 758–767.

Fettke, C. R., 1952, Tioga bentonite in Pennsylvania and adjacent states: American Association of Petrologum Geologists Bulletin, v. 36, p. 2038–2040.

Folinsbee, R. E., Baadsgaard, H., and Lipson, J., 1961, Potassium-argon dates of Upper Cretaceous ash falls, Alberta, Canada, *in* Geochronology of rock systems: New York Academy of Science Annals, v. 91, article 2, p. 352–363.

Flowers, R. R., 1952, Lower Middle Devonian meta-bentonite in West Virginia: American Association of Petroleum Geologists Bulletin, v. 36, p. 2036–2043.

French, B. M., 1984, Impact event at the Cretaceous-Tertiary boundary; A possible site: Science, v. 226, p. 353.

Galusha, T., and Blick, J. C., 1971, Stratigraphy of the Santa Fe Group, New Mexico: American Museum of Natural History Bulletin, v. 144, 127 p.

Ganapathy, R., 1980, A major meteorite impact on the Earth 65 million years ago; Evidence from the Cretaceous-Tertiary boundary clay: Science, v. 209, p. 921–923.

Gilluly, J., 1965, Volcanism, tectonism, and plutonism in the western United States: Geological Society of America Special Paper 80, 69 p.

——, 1973, Steady plate motion and episodic orogeny and magmatism: Geological Society of America Bulletin, v. 84, p. 499–514.

Gilmore, J. S., Knight, J. D., Orth, C. J., Pillmore, C. L., and Tschudy, R. H., 1984, Trace-element patterns at a non-marine Cretaceous-Tertiary boundary: Nature, v. 307, p. 224–228.

Gilmour, I., and Anders, E., 1989, Cretaceous-Tertiary boundary event; Evidence for a short time scale: Geochimica et Cosmochimica Acta, v. 53, p. 503–511.

Glass, B. P., 1984, Tektites, *in* Pye, L. D., O'Keefe, J. A., and Frèchette, V. D., eds., Natural glasses; Proceedings of the International Conference on Glass in Planetary and Geological Phenomena: Amsterdam, North Holland Physics Publishing, p. 333–343.

——, 1987, Coesite associated with North American tektite debris in DSDP site 612 on the continental slope off New Jersey, *in* Abstracts of papers, Lunar and Planetary Science Conference 18: Houston, Texas, Lunar and Planetary Institute, p. 328–329.

Gostin, V. A., Haines, P. W., Jenkins, R.J.F., Compston, W., and Williams, I. S., 1986, Impact ejecta horizon within late Precambrian shales, Adelaide geosyncline, South Australia: Science, v. 233, p. 198–203.

Gostin, V. A., Keays, R. R., and Wallace, M. W., 1989, Iridium anomaly from the Acraman impact ejecta horizon; Impacts can produce sedimentary iridium peaks: Nature, v. 340, p. 542–544.

Gratz, A., and 5 others, 1988, Optical microscopy and TEM of shocked material from the Clear Creek, Colorado K/T "magic layer", *in* Abstracts of papers, Lunar and Planetary Science Conference 19: Houston, Texas, Lunar and Planetary Institute, p. 419–420.

Gray, M. B., and Nicklesen, R. P., 1989, Pedogenic slickensides, indicators of strain and deformation processes in redbed sequences of th Appalachian foreland: Geology, v. 17, p. 72–75.

Grieve, R.A.F., 1982, The record of impact on Earth; Implications for a major Cretaceous/Tertiary impact event, *in* Silver, L. T., and Schultz, P. H., eds., Geological implications of impacts of large asteroids and comets on the Earth: Geological Society of America Special Paper 190, p. 25–37.

——, 1987, Terrestrial impact structures: Annual Review of Earth and Planetary Sciences, v. 15, p. 245–270.

Grieve, R.A.F., and Alexopoulos, J., 1988, Microscopic planar features in quartz from Scollard Canyon, Alberta, and the Cretaceous-Tertiary boundary event: Canadian Journal of Earth Sciences, v. 25, p. 1530–1534.

Grim, R. E., and Allen, V. T., 1938, Petrology of Pennsylvanian underclays of Illinois: Geological Society of America Bulletin, v. 49, p. 1485–1514.

Hale, W. E., 1955, Geology and ground-water resources of Webster County, Iowa: Iowa Geological Survey Water Supply Bulletin 4, 257 p.

Hallam, A., 1987, End-Cretaceous mass extinction event; Argument for terrestrial causation: Science, v. 238, p. 1237–1242.

Hansen, H. J., and 5 others, 1986, Cretaceous-Tertiary boundary spherules from Denmark, New Zealand, and Spain: Geological Society of Denmark Bulletin, v. 35, p. 75–82.

Hanson, R. E., and Al-Shaieb, Z., 1980, Voluminous subalkaline silicic magmas related to intracontinental rifting in the southern Oklahoma aulacogen: Geology, v. 8, p. 180–184.

Harland, W. B., and Francis, E. H., 1971, The Phanerozoic time-scale; A supplement: Geological Society of London Special Publication 5, part 1, 336 p.

Harland, W. B., Smith, A. G., and Wilcock, B., eds., 1964, The Phanerozoic time-scale; A symposium dedicated to Professor Arthur Holmes: Supplement to the Quarterly Journal of the Geological Society of London, v. 120s, 458 p.

Hartnady, C.J.H., 1986, Amirante Basin, western Indian Ocean; Possible impact site of the Cretaceous/Tertiary extinction bolide: Geology, v. 14, p. 423–426.

Hartung, J. B., and Anderson, R. R., 1988, A compilation of information and data on the Manson impact structure: Houston, Texas, Lunar and Planetary Institute Technical Report 88–08, 32 p.

Hartung, J. B., Izett, G. A., Naeser, C. W., Kunk, M. J., and Sutter, J. F., 1986, The Manson, Iowa, impact structure and the Cretaceous-Tertiary boundary event, *in* Abstracts of papers, Lunar and Planetary Science Conference 17: Houston, Texas, Lunar and Planetary Institute, p. 31–32.

Haskin, L. A., Haskin, M. A., Frey, F. A., and Wildeman, T. R., 1968, Relative and absolute abundances of the rare earths, *in* Ahrens, L. H., ed., Origin and distribution of the elements, v. 1: Oxford, Pergamon, p. 889–911.

Hass, W. H., 1948, Upper Devonian bentonite in Tennessee: American Association of Petroleum Geologists Bulletin, v. 32, p. 816–819.

Hattin, D. E., 1975, Stratigraphy and depositional environment of Greenhorn Limestone (Upper Cretaceous) of Kansas: Kansas State Geological Survey Bulletin 209, 128 p.

Heathman, J. H., 1939, Bentonite in Wyoming: Wyoming Geological Survey Bulletin 28, 20 p.

Henderson, J. R., and Vargo, J. L., 1965, Aeromagnetic map of central Iowa: U.S. Geological Survey Geophysical Investigation Map, scale 1:316,800.

Hickey, L. J., 1984, Changes in the angiosperm flora across the Cretaceous-Tertiary boundary, *in* Berggren, W. A., and Van Couvering, J. A., eds., Catastrophes and Earth history: Princeton, New Jersey, Princeton University Press, 464 p.

Hildebrand, A. R., and Boynton, W. V., 1987, The K/T impact excavated oceanic mantle; Evidence from REE abundances, *in* Abstracts of papers, Lunar and Planetary Science Conference 18 Houston, Texas, Lunar and Planetary Institute, p. 427–428.

Hildreth, E. W., 1977, The magma chamber of the Bishop Tuff; Gradients in temperature, pressure, and composition [Ph.D. thesis]: Berkeley, University of California, 328 p.

——, 1979, The Bishop Tuff; Evidence for the origin of compositional zonation in silicic magma chambers, *in* Chapin, C. E., and Elston, W. E., eds., Ash-flow tuffs: Geological Society of America Special Paper 180, p. 43–75.

Holtzman, A. F., 1970, Gravity study of the Manson "disturbed area," Calhoun, Pocahontas, Humboldt, and Webster Counties, Iowa [M.A. thesis]: Iowa City, University of Iowa, 63 p.

Honess, C. W., 1921, The Stanley Shale: American Journal of Science, v. 1, p. 63–80.

——, 1923, Geology of the southern Ouachita Mountains of Oklahoma: Oklahoma Geological Survey Bulletin 32, 278 p.

Hoppin, R. A., and Dryden, J. E., 1958, An unusual occurrence of Precambrian crystalline rocks beneath glacial drift near Manson, Iowa: Journal of Geology, v. 66, p. 694–699.

Huddle, J. W., and Patterson, S. H., 1961, Origin of Pennsylvanian underclay and related seat rocks: Geological Society of America Bulletin, v. 72, p. 1643–1660.

Hut, P., and 7 others, 1987, Comet showers as a cause of mass extinctions: Nature, v. 329, p. 118–126.

Izett, G. A., 1968, The Miocene Troublesome Formation in Middle Park, northwest Colorado: U.S. Geological Survey Open-File Report, 42 p.

—— , 1981, Volcanic ash beds; Recorders of upper Cenozoic silicic pyroclastic volcanism in the western United States beds: Journal of Geophysical Research, v. 81, p. 10200–10222.

—— , 1982, The Bishop ash bed and some older compositionally similar ash beds in California, Nevada, and Utah: U.S. Geological Survey Open-File Report 82–582, 56 p.

—— , 1987a, Authigenic "spherules" in K-T boundary sediments at Caravaca, Spain, and Raton basin, Colorado and New Mexico, may not be impact derived: Geological Society of America Bulletin, v. 98, p. 78–86.

—— , 1987b, Size of crystals and glass shards in Lava Creek B ash and of shock-metamorphosed minerals in K-T boundary rocks provide constraint on K-T boundary impact site: Geological Society of America Abstracts with Programs, v. 19, p. 713.

—— , 1987c, The Cretaceous-Tertiary (K-T) boundary interval, Raton basin, Colorado and New Mexico, and its content of shock-metamorphosed minerals; Implications concerning the Cretaceous-Tertiary impact-extinction event: U.S. Geological Survey Open-File Report 87–606, 125 p.

—— , 1988, The Cretaceous-Tertiary boundary interval, Raton basin, Colorado and New Mexico, and its content of shock-metamorphosed minerals; Implications for the Cretaceous-Tertiary impact-extinction event [Ph.D. thesis]: Fairbanks, University of Alaska, 201 p.

Izett, G. A., and Barclay, C.S.V., 1973, Geologic map of the Kremmling Quadrangle, Grand County, Colorado: U.S. Geological Survey Geologic Quadrangle Map GQ–1115, scale 1:62,500.

Izett, G. A., and Bohor, B. F., 1986a, Microstratigraphy of continental sedimentary rocks in the Cretaceous-Tertiary boundary interval in the Western Interior of North America: Geological Society of America Abstracts with Programs, v. 18, p. 644.

—— , 1986b, Comment and Reply: Geology, v. 15, p. 90–92.

Izett, G. A., and Pillmore, C. L., 1985a, Abrupt appearance of shocked quartz grains at the Cretaceous-Tertiary boundary Raton basin, Colorado and New Mexico: Geological Society of America Abstracts with Programs, v. 17, p. 617.

—— , 1985b, Shock-metamorphic minerals at the Cretaceous-Tertiary boundary, Raton basin, Colorado and New Mexico, provide evidence for asteroid impact in continental crust: EOS Transactions of the American Geophysical Union, v. 66, p. 1149–1150.

Izett, G. A., and Wilcox, R. E., 1982, Map showing localities and inferred distributions of the Huckleberry Ridge, Mesa Falls, and Lava Creek ash beds (Pearlette family ash beds) of Pliocene and Pleistocene age in the western United States and Southern Canada: U.S. Geological Survey Miscellaneous Investigations Map I–1325, scale 1:4,000,000.

Izett, G. A., Obradovich, J. D., and Mehnert, H. H., 1988, The Bishop ash bed (middle Pleistocene) and some older (Pliocene and Pleistocene) chemically and mineralogically similar ash beds in California, Nevada, and Utah: U.S. Geological Survey Bulletin 1675, 37 p.

Jansa, L. F., and Pe-Piper, G., 1987, Identification of an underwater extraterrestrial impact crater: Nature, v. 327, p. 612–614.

Johansen, M. B., 1987, Brachiopods from the Maastrichtian-Danian boundary sequence at Nye Kløv, Jylland, Denmark: Fossils and Strata, no. 20, p. 1–57.

Johnson, K. R., and Hickey, L. J., 1988, Patterns of megafloral change across the Cretaceous-Tertiary boundary in the northern Great Plains and Rocky Mountains, *in* Global catastrophes in Earth history; An interdisciplinary conference on impacts, volcanism, and mass mortality; Abstracts presented to the topical conference: Houston, Texas, Lunar and Planetary Institute Contribution 673, p. 87.

Johnson, K. R., Nichols, D. J., Attrep, M., Jr., and Orth, C. J., 1989, High-resolution leaf-fossil record spanning the Cretaceous/Tertiary boundary: Nature, v. 340, p. 708–711.

Johnsson, M. J., and Reynolds, R. C., 1986, Clay mineralogy of shale-limestone rhythmites in the Scaglia Rossa (Turonian-Eocene), Italian Apennines: Journal of Sedimentary Petrology, v. 56, p. 501–509.

Jones, E. M., and Kodis, J. W., 1982, Atmospheric effects of large body impacts; The first few minutes, *in* Silver, L. T., and Schultz, P. H., eds., Geological implications of impacts of large asteroids and comets on the Earth: Geological Society of America Special Paper 190, p. 175–186.

Kaenel, E. de, Salis Perch-Nielsen, K. von, and Lindinger, M., 1989, The Cretaceous/Tertiary boundary in the Gurnigel flysch (Switzerland): Eclogae Geologicae Helvetiae, v. 82, p. 555–581.

Kastner, M., Asaro, F., Michel, H. V., Alvarez, W., and Alvarez, L. W., 1984, The precursor of the Cretaceous-Tertiary boundary clays at Stevns Klint, Denmark, and DSDP Hole 465A: Science, v. 226, p. 137–143.

Kauffman, E. G., 1985, Cretaceous evolution of the Western Interior Basin of the United States, *in* Pratt, L. M., Kauffman, E. G., and Zelt, F. B., eds., Fine-grained deposits and biofacies of the Western Interior seaway; Evidence of cyclic sedimentary processes; Society of Economic Paleontologists and Mineralogists Guidebook, 2nd Annual Midyear Meeting, Golden, Colorado: Society of Economic Paleontologists and Mineralogists, p. 4–13.

—— , 1987, The dynamics of marine stepwise mass extinction, *in* Lamolda, M. A., and Cearreta, A., eds., 1987, Conference on Paleontology and Evolution; Extinction Events, Abstracts, 3rd Field Conference, Bilbao, Spain, October 20–23, 1987: Bilbao, Spain, University of Leioa (Vizcaya), p. 143–174.

Kay, G. M., 1931, Stratigraphy of the Ordovician Hounsfield metabentonite: Journal of Geology, v. 39, p. 361–376.

Keller, G., 1988, Extinction, survivorship, and evolution of planktic foraminifera across the Cretaceous/Tertiary boundary at El Kef, Tunisia: Marine Micropaleontology, v. 13, p. 239–263.

—— , 1989, Extended period of extinctions across the Cretaceous/Tertiary boundary in planktonic foraminifera of continental-shelf sections; Implications for impact and volcanic theories: Geological Society of America Bulletin, v. 101, p. 1408–1419.

Kendrick, D. T., 1985, Vertical distribution of selected trace elements within the Fruitland number eight coal seam near Farmington, New Mexico [M.A. thesis]: Socorro, New Mexico Institute of Mining and Technology, 181 p.

Kent, D. V., 1981, Asteroid extinction hypothesis: Science, v. 211, p. 648–650.

Knechtel, M. M., and Patterson, S. M., 1956, Bentonite deposits in marine Cretaceous formations, Hardin district, Montana and Wyoming: U.S. Geological Survey Bulletin 1023, 116 p.

Kunk, M. J., Izett, G. A., and Sutter, J. F., 1987, $^{40}Ar/^{39}Ar$ age spectra of shocked K-feldspar suggest K-T boundary age for Manson, Iowa, impact structure: EOS Transactions of the American Geophysical Union, v. 68, p. 1514.

Kunk, M. J., Izett, G. A., Haugerud, R. A., and Sutter, J. F., 1989, ^{40}Ar-^{39}Ar dating of the Manson impact structure; A Cretaceous-Tertiary boundary candidate: Science, v. 244, p. 1521–1616.

Kyte, F. T., 1986, Accretion rate of extraterrestrial matter; An iridium profile for the last 67 Ma, *in* Abstracts of papers, Lunar and Planetary Science Conference 17: Houston, Texas, Lunar and Planetary Institute, p. 452–453.

Kyte, F. T., and Smit, J., 1986, Regional variation in spinel compositions; An important key to the Cretaceous/Tertiary event: Geology, v. 14, p. 485–487.

Kyte, F. T., Zhou, Z., and Wasson, J. T., 1980, Siderophile-enriched sediments from the Cretaceous-Tertiary boundary: Nature, v. 288, p. 651–656.

Kyte, F. T., Zhou, Z., and Wasson, J. T., 1981, Extraterrestrial siderophiles and geologic extinctions, *in* Abstracts of papers, Lunar and Planetary Science Conference 12: Houston, Texas, Lunar and Planetary Institute, p. 577–579.

Kyte, F. T., Smit, J., and Wasson, J. T., 1985, Siderophile interelement variations in the Cretaceous-Tertiary boundary sediments from Caravaca, Spain: Earth and Planetary Science Letters, v. 73, p. 183–195.

Lamolda, M. A., and Cearreta, A., eds., 1987, Conference on Paleontology and Evolution; Extinction Events, Abstracts, 3rd Field Conference, Bilbao, Spain, October 20–23, 1987: Bilbao, Spain, University of Leioa (Vizcaya), 298 p.

Laubenfels, M. W., de, 1956, Dinosaur extinction; One more hypothesis: Journal of Paleontology, v. 30, p. 207–218.

Lerbekmo, J. F., and St. Louis, R. M., 1985, The terminal Cretaceous iridium anomaly in the Red Deer Valley, Alberta, Canada: Canadian Journal of Earth Sciences, v. 23, p. 120–124.

Lerbekmo, J. F., Evans, M. E., and Baadsgaard, H., 1979, Magnetostratigraphy, biostratigraphy, and geochronology of Cretaceous-Tertiary boundary sediments, Red Deer Valley: Nature, v. 279, p. 26–30.

Lerbekmo, J. F., Sweet, A. R., and St. Louis, R. M., 1987, The relationship between the iridium anomaly and palynological floral events at three Cretaceous-Tertiary boundary localities in western Canada: Geological Society of America Bulletin, v. 99, p. 325–330.

Lichte, F. E., and 5 others, 1986, New method for the measurement of osmium isotopes applied to a New Zealand Cretaceous/Tertiary boundary shale: Nature, v. 322, p. 816–817.

Lidiak, E. G., Marvin, R. F., Thomas, H. H., and Bass, M. N., 1966, Geochronology of the mid-continent region, United States: Journal of Geophysical Research, v. 71, p. 5427–5438.

Loeblich, A. R., Jr., and Tappan, H., 1964, Foraminiferal facts, fallacies, and frontiers: Geological Society of America Bulletin, v. 75, p. 367–392.

Luck, J. M., and Turekian, K. K., 1983, Osmium-187/Osmium-186 in manganese nodules and the Cretaceous-Tertiary boundary: Science, v. 222, p. 613–615.

Maddox, D. C., 1930, Bentonite in the Ordovician near Collingwood, Ontario: Science, v. 72, p. 630.

Mapel, W. J., and Pillmore, C. L., 1963, Geology of the Newcastle area, Weston County, Wyoming: U.S. Geological Survey Bulletin 1141-N, p. N1–N85.

McHone, J. F. Nieman, R. A., Lewis, C. F., and Yates, A. M., 1989, Stishovite at the Cretaceous-Tertiary boundary, Raton, New Mexico: Science, v. 243, p. 1182–1184.

McIntyre, D. B., 1968, Impact metamorphism at Clearwater Lake, Quebec, *in* French, B. M., and Short, N. M., eds., Shock metamorphism of natural materials: Baltimore, Maryland, Mono Book Corporation, p. 363–366.

McLaren, D. J., 1970, Time, life, and boundaries: Journal of Paleontology, v. 44, p. 801–815.

McLean, D. M., 1985, Deccan traps mantle degassing in the terminal Cretaceous marine extinctions: Cretaceous Research, v. 6, p. 235–259.

Melosh, H. J., 1982, The mechanics of large meteoroid impacts in the Earth's ocean, *in* Silver, L. T., and Schultz, P. H., eds., Geological implications of impacts of large asteroids and comets on the Earth: Geological Society of America Special Paper 190, p. 121–127.

Millard, H. T., Jr., 1987, Neutron activation determination of iridium, gold, platinum, and silver in geological samples: Journal of Radioanalytical and Nuclear Chemistry, v. 113, p. 125–132.

Miser, H. D., 1920, Mississippian tuff in the Ouachita Mountain region [abs.]: Geological Society of America Bulletin, v. 31, p. 125–126.

Montanari, A., and 6 others, 1984, Spheroids at the Cretaceous-Tertiary boundary are altered impact droplets of basaltic composition: Geology, v. 11, p. 668–671.

Naeser, C. W., Briggs, N. D., Obradovich, J. D., and Izett, G. A., 1981, Geochronology of Quaternary tephra deposits, *in* Self, S., and Sparks, R.S.J., eds., Tephra studies: Dordrecht, Netherlands, Reidel, p. 13–47.

Nelson, W. A., 1921, Notes on a volcanic ash bed in the Ordovician of middle Tennessee: Tennessee Geological Survey Bulletin, v. 25, p. 46–48.

Nichols, D. J., Jarzen, D. M., Orth, C. J., and Oliver, P. Q., 1986, Palynological and iridium anomalies at Cretaceous-Tertiary boundary, south-central Saskatchewan: Science, v. 231, p. 714–717.

Nichols, D. J., Brown, J. L., Attrep, M., Jr., and Orth, C. J., 1990, A new Cretaceous-Tertiary boundary locality in the western Powder River Basin, Wyoming; Biological and geological implications: Cretaceous Research (in press).

Niem, A. R., 1977, Mississippian pyroclastic flow and ash-fall deposits in the deep-marine Ouachita flysch basin, Oklahoma and Arkansas: Geological Society of America Bulletin, v. 88, p. 49–61.

Obradovich, J. D., 1984, An overview of the measurement of geologic time and the paradox of geologic time scales: 27th International Geological Congress Proceedings, v. 1, p. 11–30.

Obradovich, J. D., and Cobban, W. A., 1975, A time scale for the Late Cretaceous of the Western Interior of North America, *in* Caldwell, W.G.E., ed., The Cretaceous System in the Western Interior of North America: Geological Association of Canada Special Paper 13, p. 31–54.

Obradovich, J. D., and Sutter, J. F., 1984, What is the age of the Cretaceous-Tertiary boundary as recognized in continental rocks?: Geological Society of America Abstracts with Programs, v. 16, p. 612.

Obradovich, J. D., Snee, L. W., and Izett, G. A., 1989, Is there more than one glassy impact layer in the late Eocene?: Geological Society of America Abstracts with Programs, v. 21, p. A134.

Odom, I. E., 1976, Microstructure, mineralogy, and chemistry of Cambrian glauconite pellets and glauconite, central U.S.A.: Clays and Clay Minerals, v. 24, p. 232–238.

Officer, C. B., and Drake, C. L., 1983, The Cretaceous-Tertiary transition: Science, v. 219, p. 1383–1390.

——, 1985, Terminal Cretaceous environmental events: Science, v. 227, p. 1161–1167.

Officer, C. B., Hallam, C. L., Drake, C. L., and Devine, J. D., 1987, Late Cretaceous and paroxysmal Cretaceous/Tertiary extinctions: Nature, v. 326, p. 143–149.

O'Keefe, J. D., and Ahrens, T. J., 1982, The interaction of the Cretaceous/Tertiary extinction bolide with the atmosphere, ocean, and solid earth, *in* Silver, L. T., and Schultz, P. H., eds., Geological implications of impacts of large asteroids and comets on the Earth: Geological Society of America Special Paper 190, p. 103–120.

——, 1988, Impact production of CO_2 by the K-T extinction bolide, and the resultant heating of the whole earth, *in* Abstracts of papers, Lunar and Planetary Science Conference 19: Houston, Texas, Lunar and Planetary Institute, p. 103–120.

——, 1989, Impact production of CO_2 by the Cretaceous/Tertiary extinction bolide and the resultant heating of the Earth: Nature, v. 338, p. 247–248.

Olmez, I., Finnegan, D. L., and Zoller, W. H., 1986, Iridium emissions from Kilauea volcano: Journal of Geophysical Research, v. 91, p. 653–664.

Olson, J. C., 1976, Geologic map of the Iris Quadrangle, Gunnison and Saguache Counties, Colorado: U.S. Geological Survey Geologic Quadrangle Map GQ-1286, scale 1:24,000.

Orth, C. J., and Attrep, M., Jr., 1988, Iridium abundance measurements across bio-event horizons in the geologic record, *in* Global catastrophes in Earth history; An interdisciplinary conference on impacts, volcanism, and mass mortality; Abstracts presented to the topical conference: Houston, Texas, Lunar and Planetary Institute Contribution 673, p. 139–140.

Orth, C. J., and 5 others, 1984, An iridium abundance anomaly at the palynological Cretaceous-Tertiary boundary in northern New Mexico: Science, v. 214, p. 1314–1343.

Orth, C. J., Gilmore, J. S., Oliver, P. Q., and Quintana, L. R., 1985, Iridium abundance patterns across extinction boundaries: Geological Society of America Abstracts with Programs, v. 17, p. 683.

Orth, C. J., Attrep, M., Jr., and Qintana, L. R., 1989, Geochemical measurements across extinction boundaries in fossil record and at terrestrial impact structures: Abstracts 28th International Geological Congress, Washington, D.C., 1989, v. 2, p. 2-553.

Palme, H., 1982, Identification of projectiles of large terrestrial impact craters and some implications for the interpretation of Ir-rich Cretaceous/Tertiary boundary layers, *in* Silver, L. T., and Schultz, P. H., eds., Geological implications of impacts of large asteroids and comets on the Earth: Geological Society of America Special Paper 190, p. 223–233.

Peterman, Z. E., Sims, P. K., Zartman, R. E., and Schulz, K. J., 1985, Middle Proterozoic uplift in the Dunbar Dome of northeastern Wisconsin, U.S.A.: Contributions to Mineralogy and Petrology, v. 91, p. 138–150.

Pettersson, H., and Rotschi, H., 1952, The nickel content of deep-sea deposits: Geochimica et Cosmochimica Acta, v. 2, p. 81–90.

Pillmore, C. L., 1976, Commercial coal beds of the Raton coal field, Colfax County, New Mexico: New Mexico Geological Society Guidebook, 27th Annual Field Conference, Vermejo Park, 1976: New Mexico Geological Society, p. 227–247.

Pillmore, C. L., and Flores, R. M., 1987, Stratigraphy and depositional environments of the Cretaceous-Tertiary boundary clay and associated rocks, Raton basin, New Mexico and Colorado, *in* Fassett, J. E., and Rigby, J. K., Jr., eds., The Cretaceous-Tertiary boundary in the San Juan and Raton basins, New Mexico and Colorado: Geological Society of America Special Paper 109, p. 111–130.

Pillmore, C. L., and Tschudy, R. H., 1983, Stratigraphy of the Cretaceous-Tertiary boundary in the southern Raton basin, New Mexico and Colorado: Geological Society of America Abstracts with Programs, v. 15, p. 308.

Pillmore, C. L., Tschudy, R. H., Orth, C. J., Gilmore, J. S., and Knight, J. D., 1984, Geologic framework of nonmarine Cretaceous-Tertiary boundary sites, Raton basin, New Mexico and Colorado: Science, v. 223, p. 1180–1183.

Pipiringos, G. N., and Imlay, R. W., 1979, Lithology and subdivisions of the Jurassic Stump Formation in southeastern Idaho and adjoining areas: U.S. Geological Survey Professional Paper 1035–C, 25 p.

Pipiringos, G. N., Hail, W. J., Jr., and Izett, G. A., 1969, The Chinle (Upper Triassic) and Sundance (Upper Jurassic) Formations in north-central Colorado: U.S. Geological Survey Bulletin 1174–N, 35 p.

Pollastro, R. M., and Pillmore, C. L., 1987, Mineralogy and petrology of the Cretaceous-Tertiary boundary clay bed and adjacent clay-rich rocks, Raton basin, New Mexico and Colorado; A comparative study: Journal of Sedimentary Petrology, v. 57, p. 456–466.

Pollastro, R. M., Pillmore, C. L., Tschudy, R. H., Orth, C. J., and Gilmore, J. S., 1983, Clay petrology of the conformable Cretaceous-Tertiary boundary interval, Raton basin, New Mexico and Colorado; 32nd Annual Clay Minerals Conference, Program and Abstracts: Clay Minerals Socioety, p. 83.

Preisinger, A., and 10 others, 1986, The Cretaceous/Tertiary boundary in the Gosau Basin, Austria: Nature, v. 322, p. 794–799.

Price, N. B., and Duff, P.M.D., 1969, Mineralogy and chemistry of tonsteins from Carboniferous sequences in Great Britain: Sedimentology, v. 13, p. 45–69.

Prinn, R. G., and Fegley, B., Jr., 1987, Bolide impacts, acid rain, and biospheric traumas at the Cretaceous-Tertiary boundary: Earth and Planetary Science Letters, v. 83, p. 1–15.

Rampino, M. R., 1982, A non-catastrophist explanation for the iridium anomaly at the Cretaceous/Tertiary boundary, *in* Silver, L. T., and Schultz, P. H., eds., Geological implications of impacts of large asteroids and comets on the Earth: Geological Society of America Special Paper 190, p. 455–460.

Rampino, M. R., and Reynolds, R. C., 1983, Clay mineralogy of the Cretaceous-Tertiary boundary clay: Science, v. 219, p. 495–498.

Rankin, D. W., Stern, T. W., Reed, J. C., Jr., and Newell, M. F., 1969, Zircon ages of felsic volcanic rocks in the upper Precambrian of the Blue Ridge, Appalachian Mountains: Science, v. 166, p. 741–744.

Raup, D. M., and Sepkoski, J. J., Jr., 1982, Mass extinctions in the marine fossil record: Science, v. 215, p. 1501–1502.

Retallack, G. J., 1976, Triassic paleosols in the upper Narrabeen group of New South Wales; Part 1, Features of the paleosols: Geological Society of Australia Journal, v. 23, p. 383–399.

—— , 1988, Field recognition of paleosols, *in* Reinhardt, J., and Sigleo, W. R., eds., Paleosols and weathering through geologic time: Geological Society of America Special Paper 216, p. 1–20.

Rigby, J. K., Jr., and 5 others, Dinosaurs from the Paleocene part of the Hell Creek Formation, McCone County, Montana: Palaios, v. 2, p. 296–302.

Robertson, P. B., Dence, M. R., and Vos, M. A., 1968, Deformation in rock-forming minerals from Canadian craters, *in* French, B. M., and Short, N. M., eds., Shock metamorphism of natural materials: Baltimore, Maryland, Mono Book Corporation, p. 433–452.

Roddy, D. J., and 5 others, 1987, Computer simulations of large asteroid impacts into oceanic and continental sites; Preliminary results on atmospheric cratering, and ejecta dynamics: International Journal of Impact Engineering, v. 5, p. 525–541.

Roen, J. B., 1980, A preliminary report on the stratigraphy of previously unreported ash-fall localities in the Appalachian basin: U.S. Geological Survey Open-File Report 80–505, 10 p.

Rose, W. I., and Chesner, C. A., 1987, Dispersal of ash in the great Toba eruption, 75 ka: Geology, v. 15, p. 913–917.

Rosenfeld, A., Flexer, A., Honigstein, A., Almogi-Labin, A., and Dvorachek, M., 1989, First report on a Cretaceous/Tertiary boundary section at Makhtesh Gadol, southern Israel: Neues Jahrbuch für Geologie Paläontologie Montschefte, no. 8, p. 474–488.

Rosenkrans, R. R., 1936, Stratigraphy of Ordovician bentonite beds in southwestern Virginia: Virginia Geological Survey Bulletin 46–I, p. 85–111.

Ross, C. S., and Smith, R. L., 1955, Water and other volatiles in volcanic glasses: American Mineralogist, v. 35, p. 906–910.

—— , 1961, Ash-flow tuffs; Their origin, geologic relations, and identification: U.S. Geological Survey Professional Paper 366, 81 p.

Ross, C. S., Miser, H. D., and Stephenson, L. W., 1928, Water-laid volcanic rocks of early upper Cretaceous age in southwestern Arkansas, southeastern Oklahoma, and northeastern Texas: U.S. Geological Survey Professional Paper 154–F, p. 175–202 [1929].

Ross, R. J., Jr., Naeser, C. W., Bergstrom, S. M., and Cressman, E. R., 1981, Bentonites in the Tyrone Limestone dated by fission tracks: Geological Society of America Abstracts with Programs, v. 13, p. 541.

Rubey, W. W., 1929, Origin of the siliceous Mowry Shale of the Black Hills region: U.S. Geological Survey Professional Paper 154–D, p. 153–170.

Russell, D. A., and Rice, R. G., 1982, K-TEC II, Cretaceous-Tertiary extinctions and possible terrestrial and extraterrestrial causes, *in* Proceedings of 1981 Workshop, Ottawa, Canada: Ottawa, Ontario, National Museum of Canada Syllogeus 39, 151 p.

Ryer, T. A., Phillips, R. E., Bohor, B. F., and Pollastro, R. M., 1980, Use of volcanic ash falls in stratigraphic studies of coal-bearing sequences; An example from the Upper Cretaceous Ferron Sandstone Member of the Mancos Shale in central Utah: Geological Society of America Bulletin, v. 91, p. 579–586.

Schmidt, R. M., and Holsapple, K. A., 1982, Estimates of crater size for large body impact; Gravity scaling results, *in* Silver, L. T., and Schultz, P. H., eds., Geological implications of impacts of large asteroids and comets on the Earth: Geological Society of America Special Paper 190, p. 93–102.

Schmitz, B., 1988, Origin of microlayering in worldwide distributed Ir-rich marine Cretaceous/Tertiary boundary clays: Geology, v. 16, p. 1068–1072.

Schopf, T.J.M., 1982, Extinction of the dinosaurs; A 1982 understanding, *in* Silver, L. T., and Schultz, P. H., eds., Geological implications of impacts of large asteroids and comets on the Earth: Geological Society of America Special Paper 190, p. 415–422.

Schultz, L. G., and Wright, J. C., 1963, Bentonite beds of unusual composition in the Carmel Formation, southwest Utah: U.S. Geological Survey Professional Paper 450–E, p. 67–82.

Schultz, P. H., and Gault, D. E., 1982, Impact ejecta dynamics in an atmosphere; Experimental results and extrapolations, *in* Silver, L. T., and Schultz, P. H., eds., Geological implications of impacts of large asteroids and comets on the Earth: Geological Society of America Special Paper 190, p. 153–174.

Seiders, V. M., 1965, Volcanic origin of flint clay in the Fire Clay coal bed, Breathitt Formation, eastern Kentucky: U.S. Geological Survey Professional Paper 525–D, p. 52–54.

Sepkoski, J. J., 1988, Periodicity of extinction; A 1988 update, *in* Global catastrophes in Earth history; An interdisciplinary conference on impacts, volcanism, and mass mortality; Abstracts presented to the topical conference: Houston, Texas, Lunar and Planetary Institute Contribution 673, p. 170–171.

Shafiqullah, M., Folinsbee, R. E., Baadsgaard, H., Cumming, G. L., and Lerbekmo, J. F., 1964, Geochronology of Cretaceous-Tertiary boundary, Alberta, Canada, *in* Sastry, M.V.A., and Bhatia, S. B., eds., Cretaceous-Tertiary boundary including volcanic activity: New Delhi, International Geological Congress, 1964, Report of the 22nd Session, part 3, section 3, p. 1–20.

Sharpton, V. L., and Schuraytz, B. C., 1989, On reported occurrences of shock-deformed clasts in the volcanic ejecta from Toba caldera: Geology, v. 17, p. 1040–1043.

Shaw, H. F., and Wasserburg, G. J., 1982, Age and provenance of the target materials for tektites and possible impactites as inferred from Sm-Nd and Rb-Sr systematics: Earth and Planetary Science Letters, v. 60, p. 155–177.

Shoemaker, E. M., 1979, Earth-crossing asteroids; Orbital classes, collision rates with Earth, and origin, *in* Gehrels, T., ed., Asteroids: Tucson, University of Arizona Press, p. 253–282.

Shoemaker, E. M., Pillmore, C. L., and Peacock, E. W., 1987, Remanent magnetization of rocks of latest Cretaceous and earliest Tertiary age from drill core at York Canyon, New Mexico, *in* Fassett, J. E., and Rigby, J. K., Jr., eds., The Cretaceous-Tertiary boundary in the San Juan and Raton basins, New Mexico and Colorado: Geological Society of America Special Paper 209, p. 131–150.

Shoemaker, E. M., Shoemaker, C. S., and Wolfe, R. F., 1988, Asteroid and comet flux in the neighborhood of the Earth, *in* Global catastrophes in Earth history; An interdisciplinary conference on impacts, volcanism, and mass mortality; Abstracts presented to the topical conference: Houston, Texas, Lunar and Planetary Institute Contribution 673, p. 174–176.

Short, N. M., 1968, Nuclear-explosion-induced microdeformation of rocks; An aid to the recognition of meteor impact structures, *in* French, B. M., and Short, N. M., eds., Shock metamorphism of natural materials: Baltimore, Maryland, Mono Book Corporation, p. 185–210.

Short, N. M., and Bunch, T. E., 1968, A worldwide inventory of features characteristic of rocks associated with presumed meteorite impact craters, *in* French, B. M., and Short, N. M., eds., Shock metamorphism of natural materials: Baltimore, Maryland, Mono Book Corporation, p. 255–266.

Sides, R. J., Bickford, M. E., Shuster, R. D., and Nusbaum, R. L., 1981, Calderas in the Precambrian terrane of the St. Francois Mountains, southeastern Missouri: Journal of Geophysical Research, v. 86, no. B11, p. 10349–10364.

Silver, L. T., and Schultz, P. H., eds., 1982, Geological implications of impacts of large asteroids and comets on the Earth: Geological Society of America Special Paper 190, 528 p.

Sims, P. K., 1985, Precambrian basement map of the midcontinent, U.S.A.: U.S. Geological Survey Open-File Report 85–0604, 16 p.

Slaughter, M., and Earley, J. W., 1965, Mineralogy and geological significance of the Mowry bentonites, Wyoming: Geological Society of America Special Paper 83, 116 p.

Sloan, R. E., Rigby, J. K., Jr., Van Valen, L. M., and Gabriel, D., 1986, Gradual dinosaur extinction and simultaneous ungulate radiation in the Hell Creek Formation: Science, v. 232, p. 629–633.

Smit, J., 1984, Evidence for worldwide microtektite strewn field at the Cretaceous/Tertiary boundary: Geological Society of America Abstracts with Programs, v. 16, p. 650.

Smit, J., and Hertogen, J., 1980, An extraterrestrial event at the Cretaceous-Tertiary boundary: Nature, v. 285, p. 198–200.

Smit, J., and Kate, W.G.H.Z. ten, 1982, Trace-element patterns at the Cretaceous-Tertiary boundary; Consequences of a large impact: Cretaceous Research, v. 3, p. 307–332.

Smit, J., and Klaver, G., 1981, Sanidine spherules at the Cretaceous-Tertiary boundary indicate a large impact event: Nature, v. 292, p. 47–49.

Smit, J., and Kyte, F. T., 1984, Siderophile-rich magnetic spheroids from the Cretaceous-Tertiary boundary in Umbria, Italy: Nature, v. 310, p. 403–405.

Smit, J., and Romein, A.J.T., 1985, A sequence of events across the Cretaceous-Tertiary boundary: Earth and Planetary Science Letters, v. 74, p. 155–170.

Smith, J., Groot, H., Jonge, R. de, and Smit, P., 1988, Impact and extinction signatures in complete Cretaceous/Tertiary (K/T) boundary sections, *in* Global catastrophes in Earth history; An interdisciplinary conference on impacts, volcanism, and mass mortality; Abstracts presented to the topical conference: Lunar and Planetary Institute Contribution 673, p. 182–183.

Smith, A. G., and Briden, J. C., 1977, Mesozoic and Cenozoic paleocontinental maps: Cambridge, England, University Press, 63 p.

Smith, J. V., and Blackwell, C. S., 1983, Nuclear magnetic resonance of silica polymorphs: Nature, v. 303, p. 223–225.

Smith, R. L., and Bailey, R. A., 1966, The Bandelier Tuff; A study of ash-flow eruption cycles from zoned magma chambers: Bulletin Volcanologique, v. 29, p. 83–104.

Smith, T. A., 1971, A seismic refraction investigation of the Manson disturbed area [M.A. thesis]: Ames, Iowa State University, 65 p.

Smyth, A. L., 1984, Pedogenesis of the Olive Hill clay bed, Breathitt Formation (Carboniferous) northeastern Kentucky [M.A. thesis]: Cincinnati, Ohio, University of Cincinnati, 205 p.

Sommer, M. A., 1977, Volatiles H_2O, CO_2, and CO in silicate melt inclusions in quartz phenocrysts from the rhyolitic Bandelier air-fall and ash-flow tuff, New Mexico: Journal of Geology, v. 85, p. 423–432.

Spears, D. A., and Kanaris-Sotiriou, R., 1979, A geochemical investigation of some British and other European tonsteins: Sedimentology, v. 26, p. 407–425.

Spears, D. A., and Rice, C. M., 1973, An upper Carboniferous tonstein of volcanic origin: Sedimentology, v. 20, p. 281–294.

Stanley, W. D., Heran, W. D., Bankey, V. L., and Abrams, G. A., 1988, Configuration and structural setting of the Manson impact structure, Iowa, from surface and airborne geophysics: EOS Transactions of the American Geophysical Union, v. 69, p. 1167.

Steiger, R. H., and Jäger, E., 1977, Subcommission on Geochronology; Convention on the use of decay constants in geo- and cosmochronology: Earth and Planetary Science Letters, v. 36, p. 359–362.

Steven, T. A., and Lipman, P. W., 1976, Calderas of the San Juan volcanic field, southwestern Colorado: U.S. Geological Survey Professional Paper 958, 35 p.

Stoddard, P. R., and Jurdy, D. M., 1988, Age estimation of the Deccan Traps from the North American apparent polar wander path: Geology, v. 16, p. 296–298.

Stöffler, D., 1971a, Progressive metamorphism and classification of shocked and brecciated crystalline rocks at impact craters: Journal of Geophysical Research, v. 76, p. 5541–5551.

——, 1971b, Coesite and stishovite in shocked crystalline rocks: Journal of Geophysical Research, v. 76, p. 5474–5488.

——, 1972, Deformation and transformation of rock-forming minerals by natural and experimental shock processes: Fortschritte der Mineralogie, v. 49, p. 50–113.

——, 1984, Glasses formed by hypervelocity impact, *in* Pye, L. D., and others, eds., Natural glasses: Amsterdam, North-Holland Physics Publishing Co., p. 465–502.

Strong, C. P., and 7 others, 1987, A new Cretaceous-Tertiary boundary site at the Flaxbourne River, New Zealand; Biostratigraphy and geochemistry: Geochimica et Cosmochimica Acta, v. 51, p. 2769–2777.

Sweet, A. R., 1988, A regional perspective on the palynofloral response to K-T boundary event(s) with emphasis on variations imposed by the effects of sedimentary facies and latitude, *in* Global catastrophes in Earth history; An interdisciplinary conference on impacts, volcanism, and mass mortality; Abstracts Presented to the Topical Conference: Houston, Texas, Lunar and Planetary Institute Contribution 673, p. 190–191.

Thein, J., 1987, A tektite layer in upper Eocene sediments of the New Jersey continental slope (Site 612, Leg 95), *in* Poag, C. W., and others, eds., Initial reports of the Deep Sea Drilling Project: Washington, D.C., U.S. Government Printing Office, v. 95, p. 565–579.

Toon, O. B., and 5 others, 1982, Evolution of an impact-generated dust cloud and its effects on the atmosphere, *in* Silver, L. T., and Schultz, P. H., eds., Geological implications of impacts of large asteroids and comets on the Earth: Geological Society of America Special Paper 190, p. 187–199.

Triplehorn, D. M., and Bohor, B. F., 1983, Goyazite in kaolinitic tuff beds of Cretaceous age near Denver, Colorado: Clays and Clay Minerals, v. 31, p. 299–304.

Triplehorn, D. M., and Finkelman, R. B., 1989, Replacement of glass shards by aluminum phosphates in a Middle Pennsylvanian tonstein from eastern Kentucky: Geological Society of America Abstracts with Programs, v. 21, p. A52.

Tschudy, R. H., 1973, The Gasbuggy core; A palynological appraisal, *in* Fassett, J. E., ed., Cretaceous and Tertiary rocks of the southern Colorado Plateau: Durango, Colorado, Four Corners Geological Society Memoir, p. 131–143.

Tschudy, R. H., and Tschudy, B. D., 1986, Extinction and survival of plant life following the Cretaceous/Tertiary boundary event, Western Interior, North America: Geology, v. 14, p. 667–670.

Tschudy, R. H., Pillmore, C. L., Orth, C. J., Gilmore, J. S., and Knight, J. D., 1984, Disruption of the terrestrial plant ecosystem at the Cretaceous-Tertiary boundary, Western Interior: Science, v. 225, p. 1030–1032.

Turekian, K. K., and Wedepohl, K. H., 1961, Distribution of the elements in some major rock units of the Earth's crust: Geological Society of America Bulletin, v. 72, p. 175–192.

Urey, H. C., 1973, Cometary collisions and geological periods: Nature, v. 242, p. 32–33.

Vance, J. A., 1965, Zoning in plagioclase; Patchy zoning: Journal of Geology, v. 73, p. 636–651.

Van Schmus, W. R., and 5 others, 1987, Quimby scientific drill hole; Definition of Precambrian crustal structure in northwestern Iowa: Geological Society of America Abstracts with Programs, v. 19, p. 877.

Wackerle, J., 1962, Shock-wave compression in quartz: Journal of Applied Physics, v. 33, p. 922–937.

Wallace, M. A., Gostin, V. A., and Keays, R. R., 1990, Acraman impact ejecta and host shales; Evidence for low-temperature mobilization of iridium and other paltinoids: Geology, v. 18, p. 132–135.

Ward, P. D., and MacLeod, K., 1988, Macrofossil extinction patterns at Bay of Biscay Cretaceous-Tertiary boundary sections, *in* Global catastrophes in Earth history; An interdisciplinary conference on impacts, volcanism, and mass mortality; Abstracts Presented to the Topical Conference: Lunar and Planetary Institute Contribution 673, p. 206–207.

Watson, E. B., 1976, Glass inclusions as samples of early magmatic liquid; Determinative method and application to a South Atlantic basalt: Journal of Volcanology and Geothermal Research, v. 1, p. 73–84.

Westland, A. D., 1981, Inorganic chemistry of the platinum-group elements: Canadian Institute of Mining and Metallurgy Special Volume 23, p. 5–18.

Wilcox, R. E., 1959, Use of the spindle stage for the determination of principal indices of refraction of crystal fragments: American Mineralogist, v. 44, p. 1272–1293.

——, 1983, Refractive index determination using the central focal masking technique with dispersion colors: American Mineralogist, v. 68, p. 1226–1236.

Williamson, I. A., 1970, Tonsteins; Their nature, origins, and uses: Mining Magazine, v. 122, p. 119–124, 203–211.

Wilson, C. W., 1949, Pre-Chattanooga stratigraphy in central Tennessee: State of Tennessee Department of Conservation Division of Geology Bulletin 56, 407 p.

Wolbach, W. S., Lewis, R. S., and Anders, E., 1985, Cretaceous extinctions; Evidence for wildfires and search for meteoritic material: Science, v. 230, p. 167–170.

Wolbach, W. S., Gilmour, I., Anders, E., Orth, C. J., and Brooks, R. R., 1988, Global fire at the Cretaceous-Tertiary boundary: Nature, v. 334, p. 665–669.

Wolfe, J. A., and Izett, G. A., 1987, A new Cretaceous-Tertiary boundary section near Teapot Dome, western Powder River Basin, Wyoming: EOS Transactions of the American Geophysical Union, v. 68, p. 1344.

Wrucke, C. T., and Dings, M. G., 1979, Geologic map of the Cameron Mountain Quadrangle, Chaffee, Fremont, and Park Counties, Colorado: U.S. Geological Survey Open-File Report 79-660, scale 1:62,500.

Xie, X., and Chao, E.C.T., 1985, The variation range of optical constants and distribution characteristics of shock lamellae in shock-metamorphosed quartz: Geochemistry, v. 4, p. 97–113.

Yang, W.-H., and 5 others, 1986, Detection of high-pressure silica polymorphs in whole-rock samples from Meteor Crater, Arizona, impact sample using solid-state silicon-29 nuclear magnetic resonance spectroscopy: Meteoritics, v. 21, p. 117–124.

Zielinski, R. A., 1982, The mobility of uranium and other elements during alteration of rhyolite ash to montmorillonite; A case study in the Troublesome Formation, Colorado, U.S.A.: Chemical Geology, v. 35, p. 185–204.

——, 1985, Element mobility during alteration of silicic ash to kaolinite; A study of tonstein: Sedimentology, v. 32, p. 567–579.

Zoller, W. H., Parrington, J. R., and Kotra, J.M.P., 1983, Iridium enrichment in airborne particles from Kilauea volcano: Science, v. 222, p. 1118–1121.

MANUSCRIPT ACCEPTED BY THE SOCIETY JANUARY 3, 1990

Typeset by WESType Publishing Services, Inc., Boulder, Colorado
Printed in U.S.A. by JL Printing, Loveland, Colorado